SITE CARPENTRY

C. K. Austin

E & FN SPON

An Imprint of Chapman & Hall

London · New York · Tokyo · Melbourne · Madras

UK	Chapman & Hall, 2–6 Boundary Row, London SE1 8HN
USA	Chapman & Hall, 29 West 35th Street, New York NY10001
JAPAN	Chapman & Hall Japan, Thomson Publishing Japan, Hirakawacho Nemoto Building, 7F, 1-7-11 Hirakawa-cho, Chiyoda-ku, Tokyo 102
AUSTRALIA	Chapman & Hall Australia, Thomas Nelson Australia, 102 Dodds Street, South Melbourne, Victoria 3205
INDIA	Chapman & Hall India, R. Seshadri, 32 Second Main Road, CIT East, Madras 600 035

First edition 1979
Revised 1986
Reprinted 1987, 1989, 1991

© 1986, 1987, 1989, BTJ Books, 1991 E & FN SPON

Printed in Great Britain at the University Press, Cambridge

ISBN 0-419-15750-6

CONTENTS

PREFACE

The continued popularity of the book since its inception in 1979 must be an indication that it still fulfils a need in the field of site carpentry.

The bulk of the contents of the book deals with sound craftsmanship as evidenced by the correct use of materials and tools, the mechanics of structural design and geometry, widened from academic centre lines to take practical account of timber widths and thicknesses and the possibility of human error. The use of the scientific calculator helps to provide an easy overall accuracy in dimensions and, with practice almost instantaneous solutions to quite complex formulae.

These are all stable qualities or functions and are not likely to fall short of the minimum requirements of authoritative regulations, standards or codes of practice. However, these documents, when amended and particularly when replaced, may present the subject in a different way, the need for legal integrity often makes them appear to be obtuse. In a book such as this they only need to be studied in as far as they are relevent to sound craftsmanship as applied on site.

Perhaps the most sweeping, certainly the most heralded legal changes in the construction industry at this time are presented by the new Building Regulations 1985, produced by the Department of the Environment and the Welsh Office and revoking the 1976 regulations and their amendments.

The new Regulations make changes in the general rules of administration which allow the builder, having obtained planning permission, to choose whether the work is to be supervized by the local authority, an inspector appointed by the National House Building Council, or a privately approved inspector. Generally speaking the work can be commenced, before full approval has been given.

The Regulations themselves, although supported by brief notes are as it were skeletal in form and use non specific qualifying terms such as "adequate", "satisfactory", "reasonable", "sufficient", "safely" etc. If the builder takes them as read the onus is on him to prove to the satisfaction of the authority that the implied requirements have been met.

Also available are a series of approved documents dealing with all the regulations in detail, providing also pictorial illustrations and extensive tables. These if correctly used qualify the work for automatic approval. In as far as the above details are relevant to or come within the terms of reference of the book they are included. Within this context although they may be widened or relaxed to a limited amount they show little change.

Another notable development which has taken place is the replacement of the British Standard Code of Practice CP 112 parts 2 & 3 by British Standard BS 5268, The Structural use of timber. This is eventually to be in seven parts, but those which have already been published include part 2, Code of practice for permissible stress design, materials and workmanship and Part 3, Code of practice for trussed rafter roofs.

Changes in the new standards include the substitution of new symbols for some of those which have been in use in the U.K. for many years. These are in accordance with ISO 3898 published by the International Organization for Standardization. They include some greek symbols. Their names and uses together with others unchanged are given in Appendix A.

CHAPTER 1

The Basic Tool Kit

In this book the term carpentry is used in the wider sense and is meant to apply to all woodworking carried out on site, including formwork, carcasing and first and second fixing.

Although modern developments are more and more away from complicated joints and elaborate assemblies, the ultimate success of the work both for economy and quality still depends upon sound craftmanship with a systematic approach to the work plus overall precision and first-time accuracy in cutting, fitting and assembling. These qualities require sustained habits of concentration and the availability of satisfactory tools and equipment.

The tool kit

The carpenter's tool kit must be a matter of personal choice but it is best to start with a basic set of tools and add to them only when a special tool is needed. The writer considers that the following list should cover most of the tools needed for straightforward work.

1. An American pattern axe.
2. An American pattern claw hammer.
3. A No. 1 Warrington pattern hammer for small nails.
4. 150 and 300 mm all-steel try squares.
5. A spirit level with an additional bubble for plumbing.
6. A plumb bob, preferably a brass centre bob.
7. Two marking gauges, of different coloured woods for easy identification.
8. A 1000 mm fourfold boxwood rule; A 2000 mm flexible steel rule is a useful adjunct, but not a substitute.
9. A pair of carpenter's compasses, preferably with quadrant and screw adjustment.
10. A handsaw 6 points to the inch (25.4 mm, metrication has not caught up yet) and from 610 to 660 mm long.
11. A 500 mm panel saw with 10 points to the inch.
12. A 300 mm tenon saw with 14 points to the inch. The brass backed saw is more expensive but generally has better steel in the blade and can be straightened more easily if the edge loses its tension (Fig. 1).

13. A pad saw with spare blades.
14. A hack saw with spare blades.
15. A wooden jackplane; although these have gone out of fashion they are superior to steel planes for taking off heavy shavings, are not so likely to break if dropped and can be trued up when worn.
16. A steel fore plane or try plane 450, 560 or 610 mm long.
17. A steel smoothing plane.
18. A rebate plane; the wooden rebate plane with the skew mouth works very sweetly but the steel rebate plane is usually fitted with a fence and is easier to set.
19. A mallet; this can be home made with an ash handle and a head of beech or applewood. The writer had a mallet head of laburnum which gave many years of service.
20. An assortment of chisels. Firmer chisels are more robust but bevelled edged chisels are nicer to use and are better for cutting dovetail sockets, etc.
21. A ratchet brace with a wide swing and an assortment of bits. A screwdriver bit is useful for stubborn screws. A countersink is essential.
22. A cold chisel about 38 mm wide and a plugging chisel.
23. A steel cabinet scraper and hardened and highly polished steel sharpener are necessary if hardwood has to be cleaned up on site.
24. A 610 mm nail bar of hard octagonal section steel.
25. Various saw files, a flat mill saw file, a smooth rat tail file and any other files needed to keep the tools in good order.

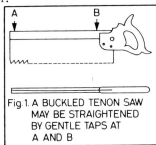

Fig. 1. A BUCKLED TENON SAW MAY BE STRAIGHTENED BY GENTLE TAPS AT A AND B

26. A combination indian oil stone and leak proof oil can.

27. A glass paper rubber purchased or home made.

28. An oily rag kept in a large flat tin.

29. A nail punch and a centre punch.

30. Several screwdrivers to suit from No. 4 to No. 12 screws.

Other equipment supplied by the contractor or made up on the job include a saw sharpening horse (most essential), various straight edges, preferably of rift sawn timber and either yellow pine or sitka spruce, a 3 m measuring rod and a 30 m steel tape. A linen tape is less open to damage but cannot be used for setting out specified measurements from drawings to a great accuracy, although it may be used for taking measurements and transferring them.

A pinch rod (Fig. 2) is useful for measuring between internal faces, e.g. in door openings.

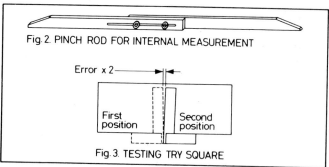

Fig. 2. PINCH ROD FOR INTERNAL MEASUREMENT

Fig. 3. TESTING TRY SQUARE

Care and maintenance of tools

The following are, in the writer's opinion, points which need to be observed in looking after and handling tools.

THE AXE: Although this should be kept sharp, the edge should be thick to ensure that the chips are forced off as they are cut.

HAMMERS: The face of the head should be polished (with old glass paper) and kept free from grease or nails will be bent over when driving. The claw hammer should not be used for drawing large nails.

TRY SQUARES: Should be checked against a straight edge (Fig. 3) and corrected or discarded if inaccurate.

SPIRIT LEVEL: The accuracy of this depends primarily upon the sensitivity of the bubble. (The bubbletube should of course have twin markings). When selecting one from several in a shop place a bus-ticket under the end of each

one on the counter in turn and buy the one which shows the greatest movement of the bubble. Test similarly for plumbing against a vertical surface. It may be tested for accuracy by reversing on a level surface. In the writer's opinion it is better to use a plumb line for testing vertical heights if great accuracy is needed.

THE PLUMB BOB: This is always positive. Using a fine line minimises the effect of the wind. A line may be plumbed single handed on to the floor (Fig. 4), a quoin may be checked both ways by a combination of sighting and measuring (Fig. 5).

FOLDING BOXWOOD RULES: Should preferably not be used for marking long lengths involving a number of pencil ticks, particularly on sawn timber. If a steel tape is not available two rules may be 'leapfrogged' (Fig. 6).

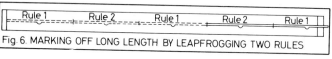

Fig. 6. MARKING OFF LONG LENGTH BY LEAPFROGGING TWO RULES

SAWS: are probably the most important part of the carpenters kit. The saw should cut freely and should not run one way or the other. When cutting to a line, half the lines should be cut away when sawing; the line of shadow along the side of the saw in the cut should remain even.

Assuming that the saw is in a bad condition, the following procedure is necessary to bring it into full working order.

1. Top off the teeth using a mill saw file in a topping clamp (Fig. 7) until the points of the lower teeth have been reached and the line of points follows a slight camber.

2. Shape the teeth and bring them all to the same size (Fig. 7). The saw file should be square to the blade and the fronts of the teeth leaning back at 15 deg. to the vertical (8 deg. for rip saws). The width of the file face should be more than twice the width of the tooth or the centre of the file will get double wear (Fig. 8).

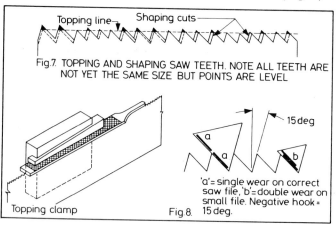

Fig. 7. TOPPING AND SHAPING SAW TEETH. NOTE ALL TEETH ARE NOT YET THE SAME SIZE BUT POINTS ARE LEVEL

'a' = single wear on correct saw file, 'b' = double wear on small file. Negative hook = 15 deg.

Topping clamp   Fig. 8.

3. Set each tooth from about half the depth of the gullet. Alternate teeth are set from each side. Too much set is not needed on clean, dry timber; but when sawing used formwork timber, this will be both gritty and wet and more set is needed. If the pliers-type set is used with a single plunger, the setting angle may be increased by grinding off the anvil and the plunger to fit.

Excess set may be removed by running an oilstone along each side of the saw, flattening off the set as required. This has the advantage of making the cut

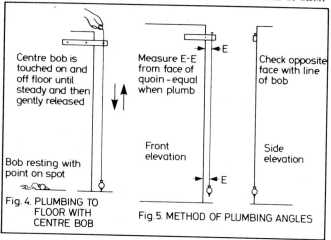

Centre bob is touched on and off floor until steady and then gently released

Bob resting with point on spot

Fig. 4. PLUMBING TO FLOOR WITH CENTRE BOB

Measure E-E from face of quoin - equal when plumb

Front elevation

Fig. 5. METHOD OF PLUMBING ANGLES

Check opposite face with line of bob

Side elevation

smoother while enabling the saw to be resharpened several times without re-setting.

4. Finally sharpen by putting the bevel on the cutting edge. The file should be held at an angle of about 70 deg. to the saw blade (85 deg. for a rip saw). Filing should be on every other gullet, the file pointing towards the handle and filing the front of the tooth leaning towards the operator. Of course this is turned round and repeated in the other gullets.

USING THE SAW: Sawing should be done without pressure using long rythmic strokes about 3/4 the length of the saw. In the case of the pad saw however that part of the saw which is to be used according to the degree of curvature of the cut should be pushed out and the cutting done in short strokes with the handle close to the timber.

PLANES: Plane irons should be shaped and the shape maintained during each sharpening as follows:

The edge of the jack plane should be rounded. The smoothing plane and try plane should have corners only rubbed out. Rebate planes should be sharpened dead straight and square with sharp corners (Fig. 9).

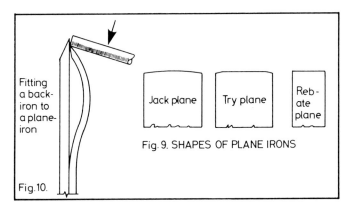

Fig.10.

Fitting a back-iron to a plane-iron

Fig. 9. SHAPES OF PLANE IRONS

Jack plane / Try plane / Rebate plane

The faces of new plane irons are generally left by the manufacturers with a milled finish which is not smooth enough to meet the bevel to give a keen cutting edge. The face of the new plane iron should be laid flat on a flat oilstone and rubbed until highly polished.

The back iron on smoothing planes, try planes, etc., should be set close to the edge. When this is done, it is sometimes found that shavings get between the two irons and choke the plane. To prevent this, the back iron may be fitted as follows:

1. Sharpen the back iron on a flat oilstone using a rocking motion until a straight wire edge is formed.
2. Turn over and remove the wire edge (to the existing under bevel as in sharpening a cutter).
3. Screw the back iron in position on the cutter and fix in the vice.
4. With the sharp corner of a round scrape sharpener, press down hard on the edge of the back iron and draw it along. The back iron will now be found to fit completely.

BRACE BITS: For screw holes, etc., spoon bits are easily sharpened, and quick cutting. For rough work and boring into end grain, the Gedge bit is robust and has a strong pull. It may be sharpened with a smooth rat tail file.

For joinery work and fitments where accuracy is needed the Russell Jennings or the Irwin bit are best. When boring holes precisely, it will be found that all bits tend to wander in the initial entry. This may be eliminated by carefully starting the hole with a centre punch.

COLD CHISELS AND PLUGGING CHISELS, ETC.: Are not hard all through but are tempered at the cutting edge. When they become worn, they should be dressed by a blacksmith and not re-ground. It is advisable to carry a spare for greater convenience.

STEEL CABINET SCRAPERS: These are simple tools but many craftsmen find difficulty in sharpening them. If however, the following procedure is adopted no further difficulty should be experienced.

1. File the edge dead square and straight, using the topping clamp if necessary.
2. Rub this edge on the oilstone (first the coarse side and then the fine), keeping it dead square, until all file marks are removed and the edge is polished.
3. Turn the scraper over flat on the stone and polish the sides until the two polished surfaces meet to form a sharp but square edge.
4. Holding the scraper upright against the body, draw the scrape sharpener hard up each edge; i.e. at an angle of about 5 deg. off the square with uniform pressure using spit as a lubricant. Only about 3 strokes are needed. (Fig. 11).

VARIOUS FILES: May be cleaned occasionally with a wire brush.

OILSTONES: Should be kept covered and clean. Fine oil or a mixture of paraffin and sump oil are satisfactory lubricants. It is important that oilstones should not be allowed to wear hollow. Narrow tools should be sharpened on them with a wandering movement.

NAIL BARS: are essentially levering tools and not chisels. A common mistake is to hammer them on the neck and weaken the most critical part. (Fig. 12).

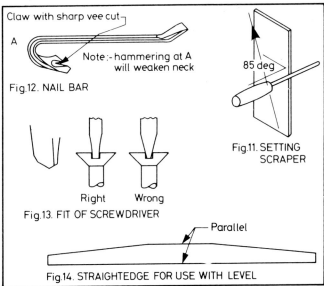

Claw with sharp vee cut

Note:- hammering at A will weaken neck

Fig.12. NAIL BAR

85 deg

Fig.11. SETTING SCRAPER

Right / Wrong
Fig.13. FIT OF SCREWDRIVER

Parallel

Fig.14. STRAIGHTEDGE FOR USE WITH LEVEL

SCREWDRIVERS: are often badly maintained, the common mistake being to taper them to a thin edge. The screwdriver should fit squarely into the slot (Fig. 13).

STRAIGHT EDGES: are tested by marking from the edge on a flat board and then turning over and checking with the original line. When a straight edge is to be used for levelling the back should be gauged off with a marking gauge (Fig. 14).

THE BASIC TOOL KIT
POWERED HAND TOOLS

It is as well to consider these tools, in the first instance, with regard to safety. This has to be taken firstly from the mechanical angle, i.e. unguarded cutters, loose parts, etc., and secondly from the direct risk from the power supply. The mechanical risks from both electric and pneumatic tools are the same and are dealt with as follows:

1. See that all necessary guards and fences are fixed and in proper working order.
2. See that cutters are sharp, securely fixed, the right way round.
3. Make sure that the job being worked on is solidly supported where this is relevant and that the line to be cut is free.

The main difference between actually using electric tools and pneumatic ones is that electric tools must be brought to top speed before loading whereas pneumatic tools must always be started under full load.

In considering the safety of pneumatic power supply, hoses should be kept short and free from kinks. All couplings should be foolproof and incapable of accidental disengagement. A loose hose jetting at 100 lbs/in² can be very dangerous. If a fault occurs with any power tool, the air supply should be cut off before investigating.

Principles of electrical equipment

The safety requirements of electric equipment are much more complex but to appreciate them it is as well to have a basic understanding of the underlying principles.

The principal risk from electric equipment is that of electric potential using the human body to escape to earth, generally by shorting through the body of the equipment itself. The human body does put up a resistance so that a low potential, say 25 to 50 volts, cannot pass and is safe. If the machine is earthed and a short occurs, a big surge of electricity passes through the earth wire, a fuse is blown and the current immediately cuts off. Without the earth wire, the human body offers a resistance, the fuse is not blown and at normal supply voltage the current continues to pass – with fatal results.

Water is a conductor of electricity and a person working in wet conditions is subject to increased risk. The risk is not only in the machine itself, but also in the supply cable at any point where it may be frayed and exposed.

Much of the electric equipment today is double insulated. If it conforms to BS 2769 in this respect and bears the BS Kitemark, it is impossible for a short into the body to occur and no earth lead is necessary. It is still possible for the cable to fray however, so the risk from the main voltage still remains.

In all electrical equipment on building sites therefore which is to be hand held, the voltage should be reduced to a safe level. The standard equipment for this is a transformer reducing to 110 volts with an earth potential (the risk to the user) of 55 volts.

It is necessary both for convenience and to avoid undue risk that electric cables should be kept as short as possible. The supply should be taken from the transformer to distribution units on each floor or in various dispersed areas, the cables being protected or kept out of harms way. From these which should be fitted with 15 amp sockets other more mobile outlet units should be taken to the areas in which the work is being done so that only short leads are needed to each machine.

Types of hand tool

Portable hand tools most likely to be of use to the site carpenter are as follows:

THE CIRCULAR SAW: This is available with from 150 to 300 mm dia. saw blades. By changing the blade it can be used to cut bricks, tiles, asbestos sheeting, metals, stone, etc. The saw can be tilted to cut at an angle of up to 45 deg to its bed and can be used against a fence or with its own fence and can be adjusted to various depths.

THE JIG SAW: This is a reciprocating saw with the blade at right angles to its bed, although it may be tilted at an angle. It will cut softwood up to 38 mm thick but its main use is cutting thin sheet material, e.g. plywood. It cuts on the up stroke.

THE SABRE SAW: This is a heavier type of reciprocating saw, the blade being in line with the forearm when held. Suitable blades can be inserted for cutting a variety of different materials.

THE DRILL: This needs no explanation. It should have two speeds for different sized bits and for cutting different materials.

THE PERCUSSION DRILL: This is fitted with a special tungsten carbide tipped impact drill on which about 160 blows are delivered per second. It is essential for the rapid drilling of concrete, granite and other hard materials.

THE SCREWDRIVER AND NUT RUNNER: This may have a screwdriver or socket wrench. The screwdriver is guided onto the screw by a hollow spring-loaded cone. A clutch (adjustable) limits the ultimate pressure put on the screw or nut.

THE PLANER: This is like an inverted panel planer with a cut up to 100 mm wide. It can be used for surface planing or narrow rebating. It is not very suitable for planing wide surfaces as the corners of the cutters tend to leave marks.

THE DISK SANDER: This carries a revolving stout rubber disk backing an abrasive disk of paper or other flexible material. It is tilted in use so that the disk only makes partial contact. It is only suitable for rough work as the action leaves circular marks and uneven surfaces.

THE ORBITAL SANDER: This carries a padded flat rectangular block to which the abrasive paper is clipped. The block does very fast circular movements of very small radius within the same rectangular area. The effect with fine abrasive is to produce a very smooth surface but with a slightly matt face.

THE BELT SANDER: This carries a continuous abrasive belt revolving on two drums at the end of a flat rectangular pad. It is faster than the other two and will produce a finish without scratches across the grain.

SITE WOODWORKING MACHINERY

This is usually confined to the normal circular saw and perhaps a small surface planer. Each machine should be treated with just as much respect as if it were in the mill. It should be protected against the weather and should preferably be securely bolted down to a temporary concrete base using, say, rawlbolts.

A very useful saw on a building site is the universal cross cut. This is basically the horizontal crosscut as is used on the mills. It can however, be raised and lowered, can be used to cut compound angles, and can also be turned at right angles (parallel to the fence) for ripping.

PRECISION IN SETTING OUT

One of the problems in working under site conditions, which often means rough, is that of securing an overall accuracy within given tolerances over a number of small consecutive measurements, without accumulating errors.

The one sure way to do this is to make all the measurements from the original point; which means some simple cumulative addition, say in the warmth of the foreman's office, before starting. The steel tape is then hooked over a nail at the start and all of the measurements taken off (Fig. 15).

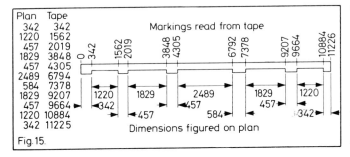

Fig. 15.

Levelling

Levels given on site from the dumpy level are not likely to be in error, but it is sometimes necessary to take intermediate levels using a spirit level and a straight edge. If this has to be done in a series of steps, the straight edge and level with it should be turned through 180 deg. at each step.

Alternate level points will then be correct even if the level or straight edge is inaccurate.

The most convenient equipment for transferring levels where corners have to be negotiated is the water level. This is basically a long rubber tube with corked glass tubes at each end filled with water so that it comes half way up the tubes (Fig. 16). A proprietary type has screw stoppers at the ends and graduated tubes.

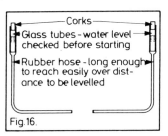

Fig. 16.

It requires two operators; one holds the glass tube against the datum and the other raises or lowers his under instruction until the water in the first tube is at the given level. The level at the other end is marked off.

There are two ways in which errors can occur:
1. If an air bubble exists in one rising end. The level at this end will then be too high.
2. If part of the rubber hose lies in the sun and gets hot.

Accuracy can easily be checked by bringing the two tubes together when the water should be level in the tubes. To avoid getting air bubbles in the tube, a large funnel with a rubber ring may be used when filling the level.

Plumbing at different storey heights

This may well be the responsibility of the engineer using a theodolite. A permanent plumb line may however be installed by hanging a heavy weight from the top storey on a steel wire in a tub of oil. When it is found to be absolutely still, it may be fixed.

CHAPTER 2

Formwork

FOUNDATIONS

One of the first jobs with which the carpenter is concerned when on site is the setting out, erection and striking of formwork to foundations. This may involve work where there are basements perhaps up to 10 m below ground. The accidents resulting from lack of appreciation of the risks due to carelessness in carrying out and maintaining the necessary excavations, still seem to continue, as evidenced by regular newspaper reports. The writer feels, therefore, that it would not be out of place to recall some of the dangers.

Timbering of trenches costs money: the temptation to leave it out, just because the ground seems solid and at the time is dry, is probably the commonest cause of accidents, resulting from the sudden collapse of the trench. There are a number of causes of failure of the apparently sound sides to excavations. Among those frequently experienced are soft intermediate layers of soil (Fig. 17), inclined strata or bedding planes (Fig. 18), or erosion by heavy rain or frost.

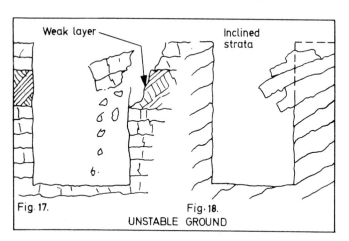

UNSTABLE GROUND

Other causes are vibration from the movement of site traffic, heavy weights placed near the edge of the trench and the side of the trench being struck, say by a concrete skip. Vehicles used to tip concrete or backfill into a trench may have to back into a blind spot. The drivers should always be aided by banksmen or signallers.

As a security against the vehicle backing too far and perhaps falling into the trench, a safety stop should be provided. One type, which may easily be constructed, consists of a timber baulk anchored at a safe distance from the edge of the trench by means of chains to well driven iron stakes, as in Fig. 19. Other proprietary types are available.

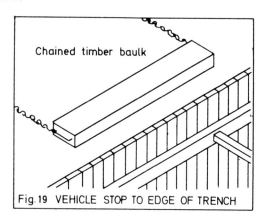

Fig. 19 VEHICLE STOP TO EDGE OF TRENCH

As an alternative to timbering of trenches in poor ground, the sides may be ramped back to the angle of repose of the soil. This means a great deal more excavation, with the temptation to reduce the working space, say outside of the ultimate position of the foundation wall. If, then, the toe of the bank is cut away (Fig. 20), this will have the effect of increasing the overall pitch of the ramp and a fall of earth will be almost certain.

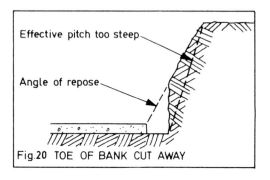

Fig.20 TOE OF BANK CUT AWAY

The Construction Regulations 1961–66, which now come within the Health and Safety at Work Act, 1974, give certain rules for safety in trenches. If these are followed explicitly, the dangers will be greatly reduced, if not eliminated. Some of them given in simple commonsense terms, in the writer's own words (no doubt full of legal loopholes) are as follows:

1. Every excavation more than 2 m deep should be inspected by a competent person before the start of each working shift. If there are special circumstances such as re-start of work after a long dry or wet spell, flooding of the excavation, or explosives have been used in the vicinity, a very thorough check should be made.
2. All timbering in trenches must be done under the direction of a competent person, using materials which have been passed for the job.
3. All struts and braces must be firmly fixed.
4. Suitable access to and egress from excavations must be provided at safe intervals (e.g. using securely lashed ladders).
5. Materials must not be stacked near the edge of the excavation.
6. All excavations more than 2 m deep must be satisfactorily protected by a guard rail. Any necessary plank bridges across the excavation must also be provided with double guard rails.

Strip foundations

In setting up formwork for foundations various types are commonly met. Strip foundations are used to carry a continuous brick, stone or concrete wall; and are wide enough to spread the load to the safe limit capacity of the soil. They may be with or without reinforcement. In firm ground, the concrete may be filled to the width of the trench and no formwork is needed. Alternatively, where the trench is wider, side forms will be needed.

When a 50 to 75 mm layer of weak blinding concrete is required to protect the steel, etc. from the mud, this can be laid easily and accurately to timber screeds. The side forms may then be butted against it and only need to be plumbed and levelled to be correct. It may be an advantage to drive a few pegs on which to rest the side forms or the forms may be made wider and the concrete levels given by means of nails suitably placed.

If the concrete blinding has already been laid, so that the forms have to rest on it, loss of grout through gaps under the forms should be prevented by strips of building paper or plastic sheeting backed up by earth or sand. Only second-

hand timber will be needed (if available) and generally speaking deflection or slight bulges in the forms under pressure from the concrete are acceptable. If, however, the concrete has to be of maximum strength or contains steel reinforcement, grout losses through bulges or gaps in the boards will cause honeycombing which will result in loss of strength and, in the presence of water, rusting of the reinforcement. It may be worthwhile to line the forms with building paper or plastic sheeting.

The composition of the side forms may depend upon the timber available and the depth of the foundation. However, as the work is being done in cramped and unpleasant conditions, it should be as simple as possible. To limit obstruction, it is better if struts, ties and distance pieces coincide at about 2 m centres; this requires some lateral stiffness in the forms. Thus, for shallow footings about 300 mm deep, 50 mm planks are best if available, as they will satisfactorily span this distance. If old 19 mm plywood has been supplied, this may be cut into strips and nailed to 100 by 50 mm walings. The sides may be supported by stakes driven into the ground.

If the soil is stony and the pegs cannot be driven straight, a cleat can be nailed to each peg as in Fig. 21. Figs. 22, 23 and 24 show forms suitable for depths of 450, 750 and 1200 mm respectively. Using 100 by 75 mm walings in each case, studs or cleats can be spaced at about 600 mm centres.

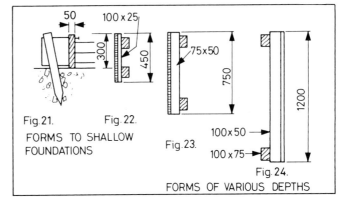

Fig. 21.   Fig. 22.
FORMS TO SHALLOW FOUNDATIONS

Fig. 23.

100 x 50
100 x 75

Fig. 24.
FORMS OF VARIOUS DEPTHS

When strutting forms off the ground or bank in soft or medium hard ground, it is essential to spread the pressures concentrated at the ends of props or struts. It is better to take a bearing on timber mud sills than nail to driven stakes. When a stake is needed at the end of a sill, as in Fig. 25; then one of rectangular section, say 150 by 38 mm, is better than a square one of, say, 75 by 75 mm.

An alternative method of strutting is to bury a 225 by 50 mm timber on edge in the ground, well rammed, and to

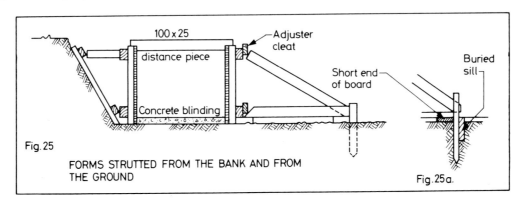

Fig. 25

100 x 25
distance piece
Adjuster cleat
Concrete blinding
Short end of board
Buried sill

FORMS STRUTTED FROM THE BANK AND FROM THE GROUND

Fig. 25a.

FORMWORK

drive the stakes in front of it, nailing the struts to the stakes (Fig. 25a).

Fig. 25 is an example of formwork to strip or beam foundation showing methods of strutting from the ground and from the bank. The cleats used behind the struts enable them to be adjusted without altering the strut lengths. The sill against the bank spreads the load from the strut and would have to be wide in soft soil. The short ends of board under the horizontal strut again spread the load.

Pile caps with foundation beams

When the sub-soil is incapable of carrying the load given, this must be taken by piles, driven through into a harder stratum. A group of these is connected together by a square or rectangular pile cap while foundation walls are taken by beams supported by the pile caps. The formwork for the pile caps will be similar to that for pad foundations (Figs. 26 and 27); while that for the beams will be identical to strip foundation forms.

In the case of the latter, it is important to realise that, although the weight of the concrete beam will be taken initially by the sub-soil, this will not be able to carry the load from the superstructure when the beam will have to live up to its name. Thus great care is needed to ensure that there is no loss of grout. If a construction joint is necessary in the length of the beam, this should always be made at the point of minimum shear which is usually in the middle third of the length.

Pad foundations or column bases

The concentrated load from floors, etc. taken by column or stanchion has to be distributed again at ground level to suit the bearing capacity of the soil. This will require a slab or block of concrete up to 3 m or more square, known as a pad or column base. If the depth has to be considerable to accommodate the punching effect of the column, a saving in concrete may be achieved, either by forming a step in the depth (Fig. 26), or splaying the sides (Fig. 27). These constructions are known as stepped pads and splayed pads respectively. The simpler pad foundation is in effect the same as the bottom part of the stepped pad so it will not be separately described.

Stepped and splayed pad formwork

This consists literally of two boxes, one on the other. The corners of the bottom box will have to be strutted from the ground or bank, as described before; but the intermediate points at the sides may be tied through the concrete.

The top part of the box may be carried by two 100 by 50 mm bearers nailed to the lower forms. If the concrete has to be poured in one, the exposed upper surface of the lower block must be cased over when its level is reached by means of panels laid against and between the bearers. These should be secured as shown by nailing to the lower forms and cleating to the bearers. If a number of holes are drilled in the top forms, these may be fixed before concreting commences. The holes will allow the air to escape and show when the required level has been reached.

In the case of the splayed pad construction the side forms are only shallow, say 225 mm, leaving the top to be formed as a truncated pyramid. The sides, therefore, may be secured by stakes, as in shallow strip foundations, but the top forms will be subject to a quite considerable upward pressure at right angles to the sloping surfaces. This pressure must be restrained, and there are several accepted methods of doing this.

If there is heavy reinforcement or a pile head, the forms may be tied down to this. Or a rough yoke of scaffold boards or similar may be nailed together, dropped over the pyramid and loaded with bags of soil, gravel or any other convenient weights. Yet a third method is to take runners along two sides of the top and nail them to driven stakes outside of the forms.

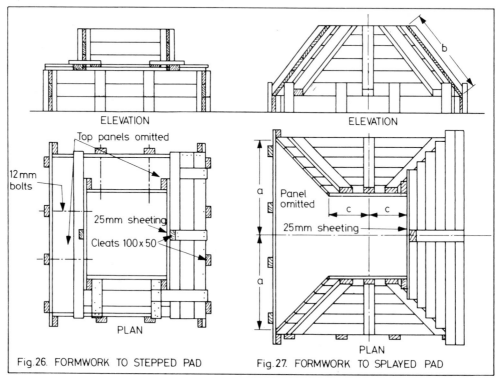

Fig. 26. FORMWORK TO STEPPED PAD   Fig. 27. FORMWORK TO SPLAYED PAD

The contract drawings will not give the true shape of the sloping sides of the pyramids which go to make the top forms. This, however, may easily be set out on, say a piece of plywood or other level surface. To obtain the true shape, use the elevation and plan details (Fig. 27) to achieve the outline shown in Fig. 28. Thus mark off 'a + a'. Set out the centre line equal to 'b'. Mark off 'c + c' parallel to 'a + a' and join up the ends to get the true shape.

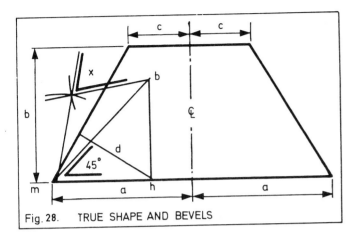

Fig. 28.   TRUE SHAPE AND BEVELS

The ends of the inner panels and the retaining cleats of the outer ones will have to be bevelled to the supplement of the dihedral angle of the corner of the pyramid. This is obtained on the already set-out true shape as follows. The line 'd' is drawn at 90 deg. to the edge of the panel. From 'm', a line is drawn at 45 deg. to the edge while from 'n' a line is drawn to intersect it at right angles. Taking the length 'd' in compasses, trammel, or radius stick, mark off from 'm' and 'b' to intersect and give the bevel required at 'x'. This may be done inwards on the panel. Although the two inner or side panels will have to be the correct shape, the outer or end panels can be any shape. However, if a number have to be made and sent on the site ready for fixing, accurate construction will pay dividends in convenience in handling.

RETAINING WALLS

In buildings planned with one or more basements, retaining walls become the most important part of the construction. There are varying procedures in casting these, related to problems in excavation. The two commonest alternatives are:

> 1. To excavate a perimeter trench around the site leaving the centre as it were an island (commonly called a 'dumpling').
> To cast the retaining walls within this trench with sufficient of the base slab to resist overturning, then, when this has gained sufficient strength, to backfill and remove the dumpling.
> 2. To excavate the whole of the site and support the bank temporarily until the retaining wall is able to take over.

Figs. 29 and 30 show one method of procedure for casting a retaining wall within a trench. The main problem here arises from the fact that the timbering necessary to support the trench is likely to obstruct the formwork necessary for

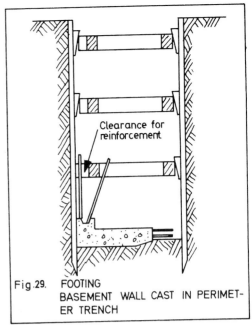

Fig. 29.   FOOTING
BASEMENT WALL CAST IN PERIMET-ER TRENCH

the casting of the wall. It is assumed that the sheet piling against what will eventually be the bank is to be left in. The reinforcing steel to the wall will have to link up with the starting bars from the foundation slab; it is better to cast a 75 mm kicker or upstand with the slab.

The screed for this can be hung from struts wedged tight between the sheeting. Room will have to be provided for the starter bars and the vertical member of the subsequent wall reinforcement. This is most easily done by blocking the walings of the timbering off the sheeting, and adjusting the blocks to accommodate the rising vertical bars (Fig. 29).

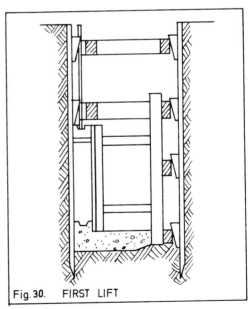

Fig. 30.   FIRST LIFT

Once the slab has been cast, the next waling up against the bank can be removed and the formwork inserted and strutted against a vertical soldier as shown in Fig. 30. This can, if necessary, be done a length at a time so that, when the concrete has been placed and initially hardened, the pressure can be taken back again through the concrete to the opposite bank. The same procedure is followed for all subsequent lifts, the second following the first as the lower part of the wall can support its part of the bank.

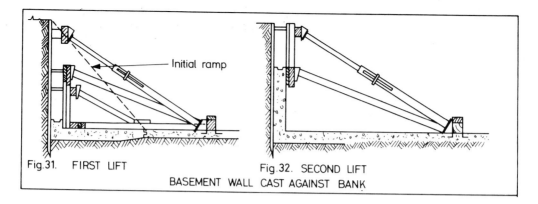

Fig.31. FIRST LIFT   Fig.32. SECOND LIFT
BASEMENT WALL CAST AGAINST BANK

Figs. 31 and 32 illustrate two stages in casting a retaining wall against a bank which has to be given interim support after the whole of the site has been excavated. The method shown, although perhaps a little involved, enables the work to be done with a minimum of excavation.

The first stage is to excavate the site without timbering, ramping the banks to the natural angle of repose of the soil. This is shown in Fig. 31, ignoring for the time the formwork and struts and taking the ramp as indicated by the dotted line. The central area of the basement slab may then be cast against suitable screeds with checkouts for a keyed joint. At the same time, mild steel wire hairpins are inserted in a line to contain a sill which is inserted when the concrete has hardened, and wedged down from the top.

The top of the bank is next supported by means of, say, adjustable steel props taken back to shaped blockings off the sill to a waling blocked off the sheeting. It is important to realise that the safe loading figures, as recommended by manufacturers of steel props, are based on the assumption that bearings at both ends will fit the end plates so that no secondary stresses (stresses other than compression) are involved. The whole of the bank may now be excavated back to the sheeting leaving the banks supported by the props.

The rest of the basement slab plus the kicker may next be cast. The screed for the kicker may be supported by stakes driven in the sub foundation and removed when the concrete has hardened sufficiently and the holes filled in.

The next procedure is a double operation. First, the formwork for the first lift is fixed and strutted back from the sill as shown. On top of this is placed a heavy timber waling, say 225 by 100 mm, faced with sheeting similar to that of the formwork panel. This is strutted by, say, 150 by 150 mm timber and adjusted by means of a wedge against a distance piece. The concrete is then filled to the top of the upper timber.

When the concrete has set hard, the first lift of formwork can be removed leaving the 150 by 150 mm strut; and the supported waling and face piece are still holding the bank through the concrete as shown in Fig. 32.

For the second (and in this case final) lift, the formwork is placed on top of the central waling and strutted back at the top as shown. The original top waling and spacers should be removed one at a time and replaced by the formwork which will then continue to support the top of the bank through the distance piece.

SETTING OUT OF FORMWORK IN FOUNDATIONS

A great deal of the work is based on centre lines. In this case, it is not necessary to fix elaborate profiles as for brick footings, etc. If a cleat is fixed to the top of a box for a pad foundation and marked with the centre line, this only needs to be aligned with the centre line of the profile. When the foundation has to take a steel stanchion, an accurate templet must be fixed to the top of the box to contain the bolts or the check-outs for these. If it is required that the bolts should be cast in with the initial pour, a sleeve, cardboard or plastic, should be placed centrally on the bolt. This, when removed, will give some allowance for adjustment in setting.

FORMWORK TERMS

To avoid confusion in the use of terms applied to formwork in different areas, the Formwork Development Group, a branch of The Concrete Society, produced a glossary of formwork terms which was adopted by the British Standards Institute and became BS4340: 1968 *Glossary of Formwork Terms*.

As some of the adopted names may not be familiar to some readers, the writer feels that it would not be out of place to give a few definitions based on the above standard. Some of these, used in this chapter but described in the writer's own words, are as follows:

ANCHOR: A device imbedded in the previously cast concrete to carry a subsequent load from formwork containing the fresh concrete (Figs. 47 and 56).

BRACE: A member, usually diagonal, which is used to stiffen the formwork rather than carry a permanent load. It may be used to plumb forms already tied against internal pressure and may be in tension or compression according to the direction of the wind or other external force (Fig. 51).

CHECK-OUT: A strip of wood used to form a recess in the concrete (Figs. 34 and 54).

DISTANCE PIECE: A short piece of timber or other material, placed between opposite forms to maintain them at a distance apart to give the required thickness of the wall, column, etc. (Fig. 36).

EDGE FORM: The board or other sheeting member which forms the side of a concrete slab in casting (Fig. 33).

FORMWORK: A word used to describe the general assembly of forms including supports. It is worth noting that the old familiar terms 'shuttering' and 'shutter' are deprecated, although they are still popular on the average site.

FORM: The framed up sheeting which contains the concrete, but not the supports. Thus a form may consist of plywood and back framing but does not include struts, braces or walings.

FORM LINING: Usually thin and comparatively weak material applied to the face of the sheeting and used either to give a smooth or textured finish to the concrete surface or to eliminate grout loss, but requiring continuous support. Examples are hardboard, thin plywood, and smooth or patterned plastic or rubber sheeting.

KICKER FRAME: The formwork used to cast a kicker or upstand to the base of a wall or column.

PANEL: A unit form such as a sheet of plywood framed in timber, being standard size to a system and interchangeable with other panels (Fig. 38).

SILL: A horizontal member resting on a solid base and taking the load from a strut or prop and transferring it to say a concrete slab. When resting on the ground, it is termed a MUDSILL.

SOLDIER: A major vertical member in a formwork system, usually outside the form and giving support to one or more forms through studs or walings.

STUD: One of a number of timbers either horizontal or vertical, which directly support the sheeting, and with the sheeting constitutes a form.

WALING: A horizontal structural timber in a formwork system. It may tie together and support a number of horizontally combined forms or panels.

WALLS

Wherever possible, it is advisable to start vertical elements in concrete with a kicker which should be 50 mm or more deep. This has two advantages:

1. The kicker which is easily and accurately formed automatically positions the base of the wall;
2. The risk of forms lifting under vibration with subsequent loss of grout is greatly reduced or eliminated.

Fig. 34 is a section through a kicker frame. The second short stake stiffens the edge form and helps the brace to support the inner member which is further stabilised by the short cleat at the top. Note that the horizontal member resting on the slab helps to prevent upsurge when the kicker frame is being filled. A cleaner finish is obtained if the inner form is added after the slab has been screeded but before it has finally set. The bevelled check-out should be fixed when the bottom of the key has been reached. Pieces will have to be short enough to thread through the reinforcement.

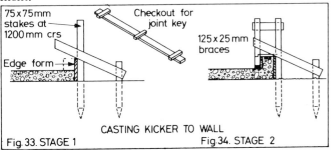

Fig. 33. STAGE 1   CASTING KICKER TO WALL   Fig. 34. STAGE 2

Before describing wall forms in more detail, it may not be amiss to discuss the basic principles of wall form design. In terms of simple mechanics, the problem starts with the internal pressure from the fluid concrete, known as 'hydrostatic pressure' (although this term strictly means 'the pressure of still water'). Its exact resolution is a complex matter and depends among other things upon the ambient temperature, rate of fill of the forms and method of vibration. It is sufficient to say, at this point, that the pressure in the forms may vary from nil at the top to 8 tonnes/m² at the lowest part of the form where the concrete is still fluid. In order to design forms for a particular job therefore, all the working conditions have to be considered and values given here are only average.

The underlying principle of all formwork design is that of collecting the overall surface pressures on the form face and transferring through a grillage of structural members to concentrated points which are then supported by ties through the concrete or external struts or props. At the opposite ends of these, the grillage principle may be applied again in reverse to spread the load over, say, the ground to the safe limit of its bearing capacity.

With the exception of struts, ties and braces, all the members throughout the grillage may be regarded as beams; each of which should, for economy's sake, be designed to span its safe maximum, giving the spacing in each case of the supporting members.

Thus, with regard to wall forms, the safe span of the sheetings gives the spacing of the studs; the safe span of the studs gives the spacing of the soldiers or walings; while the strength of these in turn governs the spacing of the struts or ties. As in all vertical forms, the hydrostatic pressure increases with depth. All horizontal members should be closer together at the bottom while vertical members should be spaced equally to accommodate the greatest pressure to be met.

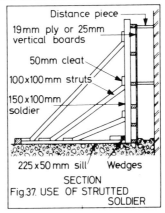

SECTION
Fig. 37. USE OF STRUTTED SOLDIER

When walls have to be cast against an existing wall or bank in one pour and some satisfactory anchorage cannot be obtained within the wall economically, the forms will have to be strutted against external pressure. One method is shown in Fig. 37 using vertical boards or plywood with the grain of the outer plies vertical. The studs are then horizontal and these are supported by strutted soldiers. The struts are driven down to tighten and backed by nailed cleats. The thrust of each set of struts is taken by a common sill as before against a continuous timber, hairpin wired to an existing concrete slab. The sills are wedged tight to the formwork; the friction against the kicker being considered sufficient to prevent form lift. It is, however, more satis-

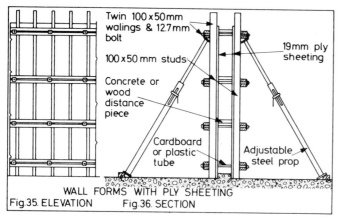

WALL FORMS WITH PLY SHEETING
Fig.35. ELEVATION   Fig.36. SECTION

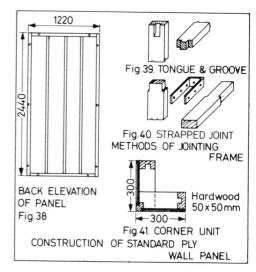

CONSTRUCTION OF STANDARD PLY
WALL PANEL

factory to provide some sort of anchor. Note that the studs are spaced closer at the bottom to allow for the increased hydrostatic pressure.

Free-standing walls

Figs. 35 and 36 show the formwork to a wall which is freestanding and is to be cast in one pour. The wall is presumed to be of limited length and conditions do not warrant the construction of special panels using plywood sheeting. The sheeting is therefore just nailed to the studs; joints being made central on a stud thickness. The internal pressure is taken by ties through the concrete. These may consist of 12 mm bolts with 100 mm square washers bearing against twin 100 by 50 mm timber walings. These are packed 25 mm apart so as to provide continuous slots to take the bolts without the need for boring holes. The bolts within the concrete area are protected by cardboard or plastic sleeves (removable after striking), which are tightened against the resistance of distance pieces to the wall thickness. If these are of timber, they are better with a wire attached taken to the top of the form to assist in removal.

Adjustable steel props are used to plumb the wall at about 3 m centres. As they do not carry any load, provision of special seatings to the props is not necessary.

This type of construction is not suitable for the multiple re-use of forms as the plywood in unprotected sheets is too vulnerable at the edges and corners. Therefore where the forms are to be used (without breaking) 'to destruction', it pays to make each set up into panels, with continuous support given to all edges of the sheet. They are likely to get some rough usage between consecutive pours; the supporting framing should therefore be strongly assembled using either tongued joints, as in Fig. 39; or butt joints with metal straps, as in Fig. 40.

The panels are butt jointed and bolted together through holes of standard spacing, semi-circular slots in the edges

coming together to provide bolt holes for the tie systems. Intermediate holes are drilled in the plywood sheeting as required. These details are shown in Figs. 42 and 42a.

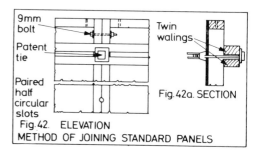

Fig.42. ELEVATION
METHOD OF JOINING STANDARD PANELS

Proprietary wall ties

Various proprietary ties are available with the general advantages that they also act as distance-pieces. When the outer parts are removed on striking, they leave the centre part in the wall with the concrete undisturbed but with clean holes which, when stopped in, protect the hidden parts from corrosion. The Rawltie (Fig. 43) has bolts passed through wooden cones into a central member with threaded ends. When the bolts and formwork are removed, the cones may be twisted out with a key.

The wedge-lock tie (Fig. 44) is tightened by a wedge with an internal keyhole slot which passes over the tie bar and tightens it by driving down through a double notch in the bar. To strike, the wedge is knocked up and removed and the form lifted off. The outer unit left in, is then screwed out taking the spacing cone with it, leaving holes to be stopped in.

Fig. 45 shows a snap tie. The forms are tightened against fixed washers. After the forms are struck, outer parts of the ties are bent over and, with the leverage thus obtained,

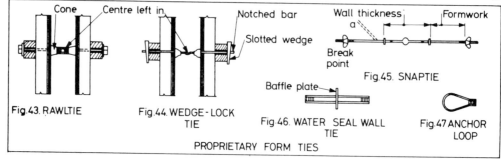

Fig.43. RAWLTIE   Fig.44. WEDGE-LOCK TIE   Fig.46. WATER SEAL WALL TIE   Fig.45. SNAPTIE   Fig.47 ANCHOR LOOP
PROPRIETARY FORM TIES

twisted off at notches 25 mm within the concrete. This leaves very small holes to be filled.

Fig. 46 shows a Rawltie incorporating a welded baffle plate. This forms an additional security against water leakage where the risk of this may be critical, say in a reservoir.

The outer angle on the return of a wall forms a weak point in the formwork as the outer walings have to cantilever the wall thickness. Tie bolts should be taken as close to the inner angles as possible. The walings both inside and outside should be locked together firmly, as shown in Fig. 48 with the pictorial detail in Fig. 49.

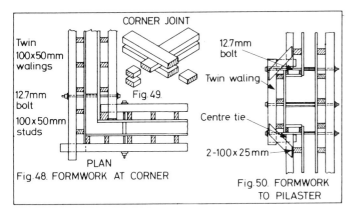

Fig. 48. FORMWORK AT CORNER

Fig. 49.

Fig. 50. FORMWORK TO PILASTER

Where a pilaster has to be formed, longer bolts will be needed. Extra pressure on the short returns will have to be countered while the angles will have to be kept square. These three points can be dealt with as shown in Fig. 50.

Climbing forms

Where the wall cannot be cast to the full height in one pour, either for reasons of excessive height, or horizontal extent; the forms will need to be lifted vertically for each successive stage. Those designed for this purpose are known as 'climbing forms'. The main problem is that of holding each lift plumb without resorting to long or elaborate struts. One way to overcome this is to use double length soldiers behind the panels, as illustrated in Figs. 51, 52 and 53.

For the first lift, after the kicker has been cast, the forms are erected and braced upright at about 2 m intervals from a stake (Fig. 51). The horizontal tie has the effect of positioning the stake from the kicker and so giving a greater rigidity to the raking brace. The soldiers, which are spaced

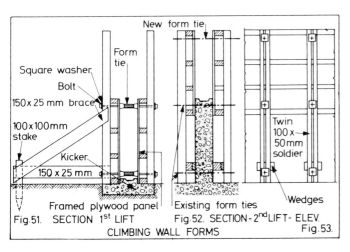

Fig. 51. SECTION 1st LIFT
CLIMBING WALL FORMS

Fig. 52. SECTION - 2nd LIFT - ELEV.
Fig. 53.

at about 1.2 m centres, are not needed at this stage and stand idle above the forms.

After the first lift, the forms are struck and raised vertically, so that the bottom bolts are now screwed to the top ties; while the soldiers, being replaced in their original position, stand with their ends below the forms as in Figs. 52 and 53. An extra set of ties is fitted to the top holes in the panels, while the folding wedges used at the bottoms of the soldiers may be adjusted to plumb the forms.

For all successive lifts, the relative positions of panels and soldiers remain the same. If a crane or other lifting tackle is available, the assembled units may be hoisted and refixed directly. If the forms have to be lifted manually, they are likely to have to be handled piecemeal. The example shown would be suitable for lifts of about 1 m, the panels being set horizontal.

Cantilevered climbing forms

When a wall has to be cast against a bank or existing wall in several lifts, then the best way to do this without excessive strutting is to use a cantilever system. This has double length soldiers bolted to anchors in the lower part of the wall, which is already cast and sufficiently hardened. The soldiers are used to support the next lift. The various stages in the operation are shown in Figs. 54 to 57, which illustrate an example of a battered wall rising off splayed footings.

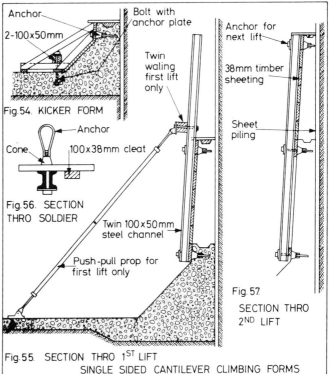

Fig. 54. KICKER FORM

Fig. 56. SECTION THRO SOLDIER

Fig. 55. SECTION THRO 1st LIFT
SINGLE SIDED CANTILEVER CLIMBING FORMS

Fig. 57. SECTION THRO 2nd LIFT

The formwork used to cast the splayed footing and kicker, which should be integral with the slab, will need anchoring down to counter the hydrostatic pressure on the sloping surface and should be fitted with anchors to take the feet of the soldiers for the first lift. These details are shown in Fig. 54.

It is assumed that each lift will be about 1.5 m so that the soldiers will be about 3 m long. When erecting the first lift, the soldiers must stand idle above the forms as in the previous example. The bottoms of the soldiers may be bolted to the anchors in the kickers, but the tops will need to be strutted. To reduce the number of struts needed, twin walings of 2/150 by 100 mm timber may be attached to the soldiers which will then only need strutting at about 1.5 m spacing.

A proprietary type of steel push-pull prop is shown bolted to the slab and also to the twin walings. As this may be adjusted to length and firmly locked, no distance pieces are needed. A further set of anchors is bolted to the tops of the forms following the procedure in the previous example.

To reduce the load as much as is economically possible, the soldiers have been shown spaced at only 600 mm centres. Under these conditions, if stout sheeting is used, intermediate studs become unnecessary. This is therefore shown as 38 mm thick solid timber made up in panels with 100 by 38 mm cleats.

To give the necessary stiffness to the soldiers without making them unnecessarily cumbersome, twin 100 by 50mm steel channels have been substituted for timber with drilled plates welded on to take the bolts. For second and all subsequent lifts as shown in Fig. 57, the walings and props may be discarded, the soldiers being kept below the forms and bolted to the anchorages in the previous pour.

As a general rule the supporting concrete should be at least 36 hours old before it can be expected to take a reasonable load. This is a matter which would have to be carefully investigated under site conditions.

COLUMNS

As for wall forms it is an advantage to cast kickers to columns at the same time as the floor or foundation concrete. The kicker frames for this should be accurately positioned and also made dead to size to ensure a good fit of the column box to the kicker subsequently formed, not only for good appearance but also to guard against grout loss. To aid this flexible foamed polyurethane adhesive strip (draught strip) should be stuck around the lid of the kicker before fixing the column box.

It is customary in modern practice to cast the column to the level of the underside of the deepest beam, strike the column forms and clamp on separate make-up forms to fit in with the ends of the beam boxes and make the intersections. Therefore, the tops of the column boxes are merely squared off to the required height. Each column box needs to be plumbed and strutted and where columns are in rows they should be carefully aligned. Continuous horizontal timbers nailed along the sides of the column boxes at the tops will help with this.

The construction of a column box depends upon its shape in plan. Column boxes are mostly rectangular and variations with straight sides such as octagonal or L-shaped in plan can be obtained, starting with a square box and blocking out to form the voids. Circular and other curved shapes are likely to be formed in sheet steel or glass fibre reinforced plastics when a number of forms with a large number of re-uses are envisaged. When only a few of these are needed, this may be done with timber, using horizontal curved ribs and narrow vertical staves to get the required outline. Other more complicated shapes may be met with in elaborately designed concrete theatres, churches and cathedrals.

14

Rectangular columns are nearly always cast using plywood sheeting, although originally solid timber using square edged or tongued and grooved boarding and timber yokes tightened one way with bolts and the other with wedges as in the column make up in Fig. 64 was used. Plywood, when used for formwork needs continuous support at the edges. As the column boxes are likely to be filled rapidly, considerable pressure will be developed at the bottom of the box and the supporting bearers will have to be closely spaced. These may be arranged in the form of frames, with horizontal rails between outer verticals, the spacing of the rails being progressively closer at the bottom to accommodate the reducing safe span of the plywood to meet increasing hydrostatic pressure, but the more usual column box construction is as shown in Fig. 63 with vertical members only.

The strength of the plywood to meet the greatest pressure will in this case govern the safe horizontal spacing between verticals (see 's' Fig. 63) and will consequently dictate whether one or more intermediates will be needed.

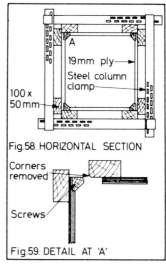

Fig.58. HORIZONTAL SECTION

Fig.59. DETAIL AT 'A'

Fig. 58 is a horizontal section through a small column box. The bearers lap the edges of the plywood, so forming checks to size, and rebates to prevent grout loss. If the corners of the intersecting members are planed off as in Figs. 59 and 62 this will help the subsequent close fits if the internal angles get slightly choked with concrete.

The corners of the column are usually moulded to a chamfer or round by means of an angle fillet (Fig. 59) or coving (Fig. 62) attached permanently to one of each of the forms at the corners of the box. They are rebated into the plywood in one panel and butt jointed to the other to avoid feather edges, which would be too fragile.

Small and medium column boxes are generally assembled with column clamps as shown in plan in Fig. 58 and in the pictorial sketch in Fig. 60. The positions of the clamps are

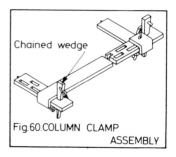

Fig.60.COLUMN CLAMP ASSEMBLY

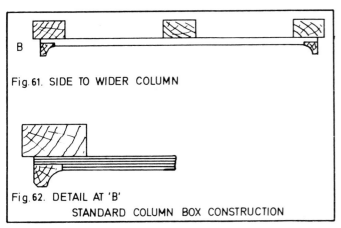

Fig. 61. SIDE TO WIDER COLUMN

Fig. 62. DETAIL AT 'B'

STANDARD COLUMN BOX CONSTRUCTION

shown in Fig. 63, their spacing 'L' depending upon the maximum safe span of the verticals. The column clamps which are invariably proprietary equipment available from steel formwork manufacturers are seldom given a specified safe bending moment. For large columns under heavy hydrostatic pressure their strength may be suspect, and may govern their maximum spacing. As a general rule it is said that when hydrostatic pressure reaches 50 kN/m^2, their spacing should not exceed 225 mm centres.

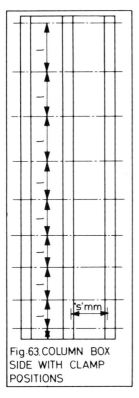

Fig. 63. COLUMN BOX SIDE WITH CLAMP POSITIONS

Fig. 64 shows how the top of the column box may be made up to take the beam intersections after the column box has been struck. The sheeting, cut out of the beam box outline is from tongued and grooved boarding but each side has a central butt joint for ease of striking. Beam sides butt against the column sheeting but the beam bottom is better carried through to the concrete face as it will be the last to be struck. Any nails not in the concrete face should be driven with the heads standing for ease of withdrawal.

Fig. 65 is a section through the formwork to a circular column. The cylindrical outline is formed with narrow vertical staves which will not need to be hollowed. The

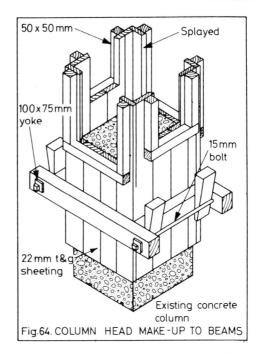

Fig. 64. COLUMN HEAD MAKE-UP TO BEAMS

staves being narrow, will be individually weak and should therefore be 50 mm thick. If the joint marks left by the staves on the concrete face are objected to (some architects like to feature them) the box can be lined with hardboard. This will bend more easily if soaked in water, but should be surface dried and oiled before fixing the formwork.

The retaining yokes, which are spaced vertically to accommodate the limiting safe spans of the staves, each consist of two semi-circular ribs built up from three thicknesses of 19 mm plywood with staggered joints. Each half

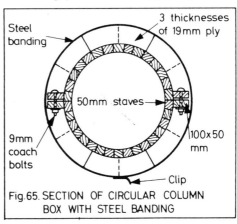

Fig. 65. SECTION OF CIRCULAR COLUMN BOX WITH STEEL BANDING

is housed and nailed into 100 by 50 mm vertical studs so that the two half forms can be bolted together in assembly. The necessary strength to resist hydrostatic pressure is given by steel bands (as used commercially in bundling sheets of plywood, etc.) which are tightened around the yokes and crimped together with a clip by means of a special tool.

It should be noted that the fluid pressure within a cylinder such as this only creates tension in the containing circular ribs. There are no other distorting stresses.

BEAM BOXES

Concrete beams cast in situ may form part of a flooring system consisting of main and secondary beams arranged

15

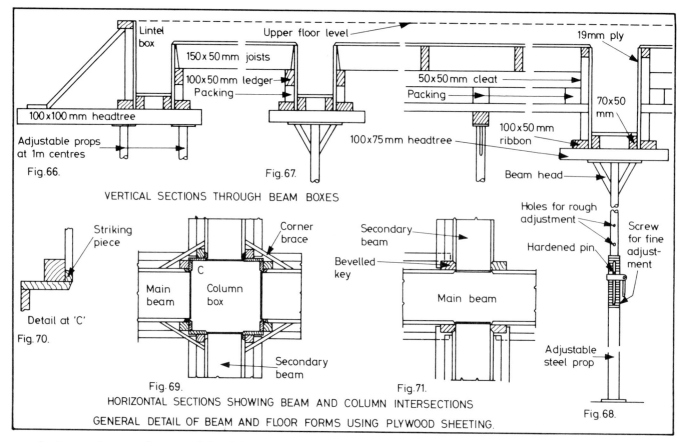

Fig.66.
Fig.67.

VERTICAL SECTIONS THROUGH BEAM BOXES

Fig. 70.

Fig. 69.

Fig.71.

Fig.68.

HORIZONTAL SECTIONS SHOWING BEAM AND COLUMN INTERSECTIONS
GENERAL DETAIL OF BEAM AND FLOOR FORMS USING PLYWOOD SHEETING.

to suit the maximum safe span of the slab, which may be cast with, or after the slab, or the beams may be isolated with the purpose of, say, carrying a wall or other super-structure.

Although initially beam boxes were formed from solid timber sheeting cleated into panels, the sheeting used today is almost invariably plywood as this gives a better surface and freedom from bad joints with subsequent grout loss. Figs. 66 to 68 are sections through the formwork to a floor with main and secondary beams using plywood sheeting. The main characteristic of plywood sheeting which influences formwork design is its lack of stiffness in either direction as compared with the longitudinal stiffness of solid timber.

To counter this, all edges of plywood panels subject to appreciable pressure should be given continuous support. The drawings illustrate how this has been carried out.

Fig. 66 shows an external secondary beam or lintel. The beam bottom is carried by 75 by 50 mm runners supported by timber headtrees on pairs of steel props with 150 mm square heads, more fully illustrated in Fig. 73.

The headtrees cantilever out to provide a basis for strutting the external beam side. Both beams are retained at the bottoms by 100 by 50 mm ribbons. The outer panel is stiffened by 50 by 50 mm cleats at about 600 mm centres while a flat 100 by 50 mm waling supports the outside at the top and provides a good line. Raking struts are taken, as shown, off each headtree.

On the inside of the beam box the decking to the floor slab is carried on 150 by 50 mm joists at about 600 mm centres. These rest on 100 by 50 mm ledgers nailed to the plywood sides and packed off the runners above the head-trees as shown. The ends of the joists are splayed about 5 degrees so that for striking, one end can be dropped

without binding. The edges of the decking panels are bevelled for the same reason and also kept in about 3 mm from the beam sides so that they will not key in under side pressure.

Fig. 67 is a section through an internal secondary beam box. As there is not the same amount of eccentric loading as on the lintel a single steel prop, but with a beam head, is used for support. It is essential that concrete should be placed evenly on both sides of the beam. Other details are as have been already described for the inside of the lintel form.

Fig. 68 is a section through a main beam form. Although the beam is larger and the beam bottom itself will be more heavily loaded, the propping system will have to carry less weight as most of the weight of the slabs is carried by the secondary beam forms. Fig. 68 also gives more details of an adjustable prop and illustrates the method of adjustment or striking. The square head in the standard prop is drilled to take a half round channel to carry a scaffold tube or a stirrup head (Fig. 72) to carry a runner.

The make-up of column forms has already been dealt with, but the plan in Fig. 69 shows how the beam boxes are fitted into this. Some provision must be made for striking, and Fig. 70 shows how a striking piece nailed to the 50 by 50 mm cleat gives clearance for the beam side when removed. To resist pressure and prevent spreading of the forms at the top of the beams corner braces are nailed to the beam sides as shown.

Fig. 71 shows the intersection between the main and secondary beam forms. An alternative aid to striking using bevelled keys is also illustrated. The secondary beam bottoms which will be the last to be struck run over the main beam sides to the concrete face and are propped underneath by packings as shown in Fig. 68.

In small beams from 450 to 600 mm deep there will be little hydrostatic pressure on the beam sides at the top. In the secondary beams these are supported by the ends of the joists, while the main beam sides are supported by the intersecting secondary beam sides plus an occasional nail into the stiffened tops.

As an alternative to casting all the beams and slabs together, the beams may be cast first to the underside of the slab level and the decking erected afterwards. This simplifies the formwork to a certain extent. Following the casting of the beams, the beam bottoms and supports may be left in and the ends of the joists propped off them, or the whole of the beam formwork may be struck and the decking then set up and propped as a separate procedure.

In these cases the side forms to the beams will have to be strutted externally as in Fig. 66 or tied across from beam to beam. Small beam boxes may be tied across the tops to stop them from spreading or column clamps may be clamped around them. If beams are of greater depths then more intense hydrostatic pressures have to be guarded against and tie bolts will have to be used at the top and bottom of the forms.

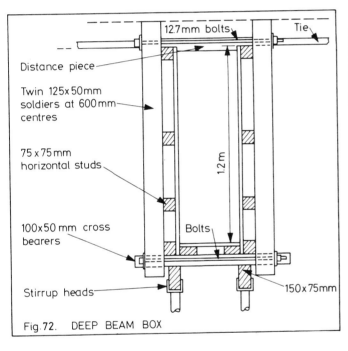

Fig. 72. DEEP BEAM BOX

Fig. 72 is a section through a beam box 1200 mm deep concrete size. It is treated as an isolated beam or one which has to be cast before the slab. To accommodate the considerable dead weight of the beam instead of headtrees, 150 by 75 longitudinal runners are used supported by stirrup heads on steel props or some other proprietary system. This enables the bearers under the beam bottom to be close spaced. Beam sides are made up with plywood panels with horizontal studs closer at the bottom to take increasing hydrostatic pressure.

The lateral support is given by twin soldiers bolted through above and below the concrete and spaced at about 600 mm centres. The tops of the beams are tied back to other beams or some other convenient anchorage. If the slab had to be cast with the beam then the tops of the soldiers would be kept below the soffit line and the top bolt taken through the concrete.

SUSPENDED SLABS

This term covers horizontal slabs above ground in floors, roofs, canopies, etc. The primary need is to erect a level and smooth leak-proof decking of sufficient strength to carry the concrete, and any superimposed loads, without failure or undue deflection, and to concentrate this on to whatever propping system is being used. Timber props adjusted by means of folding wedges, at one time universally popular for this purpose, have now been replaced by adjustable steel props, or when the height exceeds about 5 metres, steel scaffolding or various other proprietary propping systems.

For the maximum economy of materials units should be used to their maximum safe capacity which generally means to the widest spacing or greatest span.

The common terms for slab forms are:
Decking—the sheet material on which the concrete rests, usually of plywood.
Joists—carrying the decking, ledgers carrying the joists, and props carrying the ledgers. The joists are spaced to suit the safe span of the sheeting. The ledgers are spaced to suit the safe span of the joists while the props are spaced to suit the safe span of the ledgers.

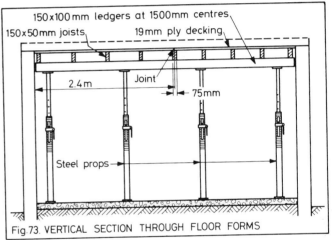

Fig. 73. VERTICAL SECTION THROUGH FLOOR FORMS

Fig. 73 shows a vertical section through the formwork to a concrete slab. As it is within four walls, these restrain the forms against any lateral movement. If, however, the slab is wholly or partly free-standing the props will need to be braced, usually by means of scaffold tubes.

The plywood is laid directly on the joists with an occasional small nail to hold it in place. As more than one standard sheet is needed for length and width, no special

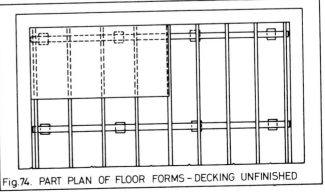

Fig. 74. PART PLAN OF FLOOR FORMS - DECKING UNFINISHED

striking provisions are needed. If striking of decking is likely to be difficult, two joists may be placed close together at one side of the slab to accommodate a separate striking piece bevelled for easy removal. Fig. 74 is a part plan showing the layout of ledgers, joists and props.

It is very important to ensure that all props are truly upright when finally setting them. Fig. 75 shows the effect of the load on a prop which is slightly tilted. As the head and base plates are made square to the axis of the prop the load and reaction bear on the points 'P' indicated by the arrow.

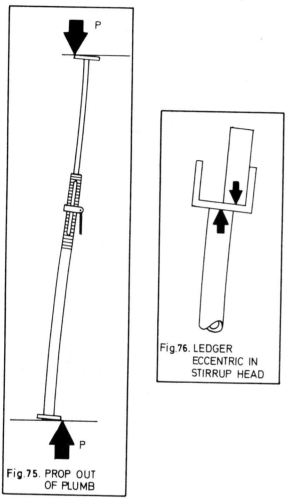

Fig.75. PROP OUT OF PLUMB

Fig.76. LEDGER ECCENTRIC IN STIRRUP HEAD

This creates secondary stresses in the shaft and will cause buckling as shown. Once a prop or strut starts to buckle or bend, the bending of itself increases the bending moment without any further load being applied and there will come a point, quite early, when the buckling will continue without any further load to the point of collapse.

The same thing can occur when a prop or stirrup head is eccentrically loaded as shown in Fig. 76. The supported ledger should never be more than 25 mm out of centre. A stirrup head can always be centralised on a supported timber by twisting the head to a maximum until it is diagonal to the timber on plan.

The fixing of an edge form to a slab well above ground level can sometimes be a problem. One way to deal with this is to wire through to the decking on the inside. If the supporting wall is also in concrete, bolt holes or proprietary wall tie centres can be left in the top of the wall and both inner ledger and edge forms can be bolted through or into the wall.

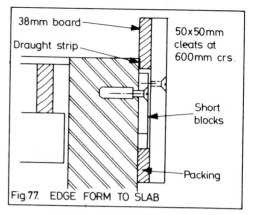

Fig.77. EDGE FORM TO SLAB

When the wall is fair faced brickwork, not to be defaced, then assuming a power drill is available the method shown in Fig. 77 may be used. A series of holes are drilled into the joints in the brickwork at convenient levels and spacings and rawlplugs or similar proprietary fixings inserted.

A separate wooden block slightly thinner than the edge board is screwed to each plug as shown. No precision is needed. Short cleats carrying the edge board and suitably packed out are screwed to these at the correct level. A draught strip stuck to the lip of the wall prevents loss of fines at the joint.

Fig. 78 shows a method of supporting the ends of the joists to a slab by means of hangers, thus saving on the number of props needed. Metal hangers are hooked into shallow slots left in the tops of the walls and covered with polystyrene blocks as shown. If the hangers are kept about 1 metre centres smaller ledgers may be used. Final adjustment is by means of folding wedges.

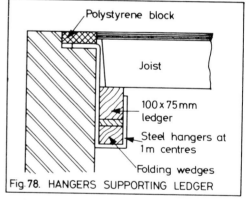

Fig.78. HANGERS SUPPORTING LEDGER

Although the simplest way of sheeting slab formwork is by means of standard plywood sheets laid direct on the joists, the edges and particularly the corners of the plywood are very vulnerable. Appreciable damage to edges and corners may mean replacement of a sheet if many re-uses are planned, and it may be advisable to make up the decking into framed panels incorporating the joists, with additional timbers at the ends. Fig. 79 shows details of a method of finishing decking panels for a maximum number of re-uses without lowering the standard of finish.

Using film-faced birch plywood which is impregnated with protective resin for maximum wear, this is backed by 100 by 75 mm intermediate joists with 100 by 50 mm timbers at the side which only carry half loads and at the ends. The whole panel is edged with hardwood strips, the narrow contact surfaces of which ensure a closer fit on the surface. The joists between panels can finally be sealed with

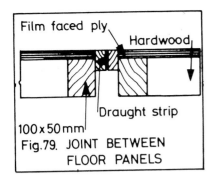

Fig.79. JOINT BETWEEN
FLOOR PANELS

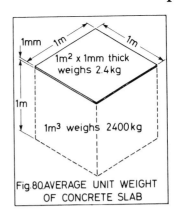

Fig.80.AVERAGE UNIT WEIGHT
OF CONCRETE SLAB

adhesive draught stripping. If the hardwood edges get damaged they can easily be replaced. If the plywood is screwed to the back framing (using temporary stopping to cover the screwheads) the screws can be removed and the panels reversed to give double life without loss of quality.

BASIC DESIGN CALCULATIONS

The high cost of timber and other materials today emphasises the need to reduce the amount used, particularly for temporary work such as formwork where they add nothing to the final structure.

It must not be forgotten that the finished concrete is only as good and well shaped as the formwork which held it. It is a false economy which, through skimping the sizes or number of supports leads to unsightly bulges or failure of the concrete through unacceptable deflection or collapse of the forms.

To keep the line of economic success between extravagance on the one hand and dangerous weakness on the other, the formwork designer needs to know the magnitude and distribution of the loads which the formwork will have to carry, the stresses these produce in the structural materials used, and their strength to resist them.

From this information he must calculate the minimum safe section of supporting units and their maximum safe spacing.

Loads and pressures on formwork may be separated into two well-defined sections. These are:

1. Dead loads on horizontal forms or decking from the concrete, plus imposed loads such as live loads from workmen, barrows, plant, etc., and bulking of concrete before spreading and
2. Hydrostatic pressure on vertical forms in walls, etc., due to the fluid state of the unset mix plus impact loads from concrete dropped from a height.

In each case the self weight of the timber formwork (the carpenter's main concern) is generally ignored.

Dealing first with the design of horizontal forms as for floors, flat roofs, etc., the dead weight of course depends upon the density of the concrete, this will vary according to the nature of the mix and the kind of constituent aggregate, but for purposes of calculating loads is most conveniently taken in Newtons at 24000 N/m³.

For the purposes of design it is necessary to know the pressure per m² per mm thickness exerted upon the decking. This may be taken as $\frac{24000}{1000}$ or 24 N/m² (Fig. 80).

The imposed load cannot be so accurately calculated, but values based on experience are 3500 N/m² for loads on decking and 2000 N/m² for loads on structural members where the local load imposed by site activity must represent a smaller percentage increase to the overall dead load which the member is designed to carry.

Fig. 81 shows diagrammatically the unit design load on decking from a 150 mm thick slab. Thus the weight of concrete

$$
\begin{aligned}
&= 24 \times 150 &&= 3600 \text{ N/m}^2 \\
\text{Imposed load} &&&= 3500 \text{ N/m}^2 \\
\hline
\text{Total load} &&&= 7100 \text{ N/m}^2
\end{aligned}
$$

If any special loads are likely to be applied to the decking such as from the storage of hollow tiles or reinforcement or the excessive heaping of concrete before spreading, these items must be especially allowed for.

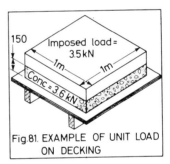

Fig.81. EXAMPLE OF UNIT LOAD
ON DECKING

It will be seen from the diagram including Figs. 82 to 85 that the structure of formwork is so arranged that uniformly distributed loads from the concrete are progressively

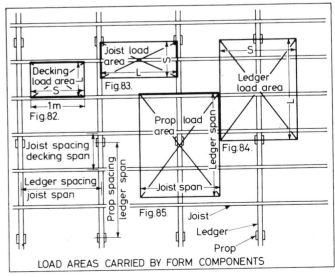

LOAD AREAS CARRIED BY FORM COMPONENTS

gathered and finally concentrated on to central supports which in the case of horizontal forms are the props, usually telescopic proprietary steel units.

Loads from the decking (Fig. 82) are taken by the joists, and loads from the joists are taken by the ledgers (Fig. 83) whilst loads from the ledgers are taken by the props (Fig. 84).

As the loads in each case depend upon the area of concrete carried it is convenient to give the dimensions of this as factors of the total loads represented by 'F' in the basic equations. Thus if f is taken as the unit load per square metre, s is taken as the width of the load in metres or part of a metre and L is taken as the length of the load (identical with the span) in metres then design load in each case = fsL.

However as all equations relating to design will be taken in units of Newtons and millimetres s and L must be expressed in mm.

$$\text{So F becomes } \frac{fsL}{10^6}.$$

The main advantage of analysing 'F' in this way is that variations in loading, and subsequent stresses due to alteration in load areas can be dealt with direct without trial and error.

Calculations relating to decking

Sheeting used in forming decking is nearly always plywood so examples will be confined to plywood in this article. If the plywood is in standard sheets the spacing of the supporting joists, will, to be economical, be a factor of the length of 8 ft., say 2 ft. or 600 mm or 19.2 in. or 480 mm.

In the case of the decking the value of 's' for the width of the load must be the same as width 'b' for the bending or deflection formula and to coincide with 'Z and I' values given in tables must be 1000 mm.

The design of plywood in decking is usually governed by deflection limits while for joists and ledgers which are more heavily loaded in relation to value 'b' the criterion is bending strength.

Plywood is of complex construction the plies at right angles to the span do little to increase the bending strength and are usually ignored. Therefore where possible the outer plies should be parallel to the span and this condition will be assumed in the following calculations.

Various development bodies concerned with the expansion of the plywood trade produce booklets with tables of values. In this article relevant data is taken from the booklet produced by the Council of Forest Industries of British Columbia and relates to Douglas fir plywood.

Values consulted in tables include safe fibre stress $\sigma_{m,adm,\parallel}K_3K_{16}$. See appendix A. To avoid the tedium of repetition of subscripts and modification factors it is recommended that these be worked out at an early stage and then the ultimate calculated stress be represented by a single letter e.g. σ. Thus taking the dry grade stresses of selected plywood in bending with the outer plies parallel to the span and with modification factors K_3 for short term loading and K_{36} for wet exposure of plywood we can calculate ultimate stress value $= 11.6 \times 1.5 \times 0.7 = 12.2 \text{ N/mm}^2$ so then this can be registered as $\sigma = 12.2 \text{ N/mm}^2$. Modulus of elasticity E. The section modulus (for 1 m wide) Z and the moment of inertia (for 1 m wide) I. To find the minimum required thickness of plywood either Z or I are calculated and then checked against relevant thicknesses of plywood in the tables.

EXAMPLE 1: Given that the concrete slab is 150 mm thick and the spacing of the joists 600 mm, to calculate a suitable thickness for the plywood decking. (1) for strength in bending.

From the previous example as in Fig. 81.

$$f = 7100/m^2$$

$$F = \frac{fsL}{10^6}$$

$$s = 1000 \quad L = 600 \text{ mm}.$$

As the plywood will span several supports bending moment

$$= \frac{FL}{10}.$$

Bending moment equation is

$$\frac{FL}{10} = \sigma Z.$$

By substitution

$$\frac{fsL^2}{10 \times 10^6} = \sigma Z.$$

Therefore

$$Z = \frac{fsL^2}{10^7 \sigma} = \frac{7100 \times 1000 \times 600^2}{10^7 \times 12.2} = 20950 \text{ mm}^3.$$

From the tables 16 mm 5-ply gives z = 21000 so that this would be satisfactory for bending.

EXAMPLE 2: To check for deflection.

Equation for deflection is

$$D = \frac{3}{384} \times \frac{FL^3}{E.I}.$$

For deflection of decking the limit is generally

$$D = \frac{L}{270}.$$

From the tables

$$E = 11031 \text{ N/mm}^2.$$

It is required to find the value of I
By substitution,

$$\frac{L}{270} = \frac{3}{384} \frac{fsL}{10^6 E I}.$$

By transposition

$$I = \frac{270 \times 3fsL^3}{384 \times 10^6 nE}.$$

By substitution

$$I = \frac{270 \times 3 \times 7100 \times 1000 \times 600^3}{384 \times 10^6 \times 11031} = 293259 \text{ mm}^4.$$

From the tables 19 mm 7 ply gives I = 311000 so 19 mm ply would be needed and deflection is the criterion.

Joist design

Assuming that joists are at a given spacing (say 600 mm crs) (Fig. 83) then there may be two alternatives, either the span may be given and the minimum safe size of the joist calculated or the size of the joist given and the maximum span calculated.

Taking M first. It is safer to assume that the joists will be simply supported, i.e. to span only between two ledgers.

The M is therefore $\dfrac{FL}{8}$ and the bending moment formula

$$\frac{FL}{8} = \frac{\sigma bh^2}{6} \quad \text{From CP112/1971.}$$

Taking $\sigma_{m,adm,\|}K_2K_3$ for bending, wet exposure and duration of loading (Tables 14 and 15 in BS 5268) (obviously parallel to the grain) and multiplying out we get $7.4\,\text{N/mm}^2$. We can then simply state $\sigma = 7.4\,\text{N/mm}^2$ for bending only. The imposed load can be reduced to $2000\,\text{N/m}^2$ so $f = 2000 + 3600 = 5600\,\text{N/m}^2$.

$$\frac{FL}{8} \text{ becomes } \frac{fsL^2}{8 \times 10^4} = \frac{\sigma bh^2}{6}.$$

EXAMPLE: 3: To calculate the minimum safe size of the joist given the span = 1500 mm.
Then by transposition

$$bh^2 = \frac{6fsL^2}{8 \times 10^6 \sigma}.$$

By substitution

$$bh^2 = \frac{6 \times 5600 \times 600 \times 1500^2}{8 \times 10^6 \times 7.4} = 766216\,\text{mm}^3.$$

A choice of joist thickness is now given. Assuming b = 50 mm then

$$h = \sqrt{\frac{766216}{50}} = 124\,\text{mm}$$

so 125×50 mm joist would be suitable.

EXAMPLE 4: To check for deflection. Deflection formula for a simply supported beam is

$$D = \frac{5}{384} \frac{FL^3}{EI}.$$

From BS 5268 part 2 table 8 for strength grade SC3, $E_{mean} = 8800$ multiplied by 0.8 (table 14) for wet conditions, $= 7040\,\text{N/mm}^2$ I (for square timber beam)

$$= \frac{bh^3}{12} \Bigg| D = \frac{L}{360}.$$

By substitution

$$\frac{L}{360} = \frac{5}{384} \times \frac{12fsL^4}{10^6 \times Ebh^3}.$$

By transposition and cancellation

$$bh^3 = \frac{360 \times 5 \times 12fsL^3}{384 \times 10^6 \times E}.$$

By substitution

$$bh^3 = \frac{360 \times 5 \times 12 \times 5600 \times 600 \times 1500^3}{384 \times 10^6 \times 7040}.$$

Let b = 50 mm then

$$h = \sqrt[3]{\frac{90607244}{50}} = 121.9\,\text{mm}$$

so 122×50 mm would be suitable for deflection and bending governs the size.

EXAMPLE 5: Given size of joist to calculate maximum safe span, given other conditions as before, with 125×50 mm joists.
Taking the formula

$$\frac{fsL^2}{8 \times 10^6} = \frac{\sigma bh^2}{6} \quad \text{when } \sigma \text{ stands for } \sigma_{m,adm,\|}K_2K_3$$

and giving in terms of L.

$$L^2 = \frac{8 \times 10^2 \sigma bh^2}{6fs}.$$

By substitution

$$L = \sqrt{\frac{8 \times 10^6 \times 7.4 \times 50 \times 125^2}{6 \times 5600 \times 600}} = 1515\,\text{mm}.$$

Note. The slight variation from the previous example for L is due to rounding off value of d.

Design of ledgers

These again are best considered as simply supported. The loading however is not uniformly distributed but is imposed at a number of points the distance apart being equal to the spacing of the joists.

However, if the sizes of the ledgers are known and their maximum safe span (spacing of the props (Fig. 84)) is required, a very close values can be obtained by taking the overall area of concrete carried and treating as a uniformly distributed load.

The precise maximum bending moment will depend partly upon the actual positions of the joists on the ledger; as this is not likely to be known the position of the joists for the maximum bending moment is taken.

This is when one joist is at mid span and the positions of the others spaced from it.

In the following example, the same relevant conditions as before are taken the approximate span of the ledger being obtained by the formula

$$L = \frac{8 \times 10 \times \sigma bh^2}{6fs}.$$

This will enable the number of joists carried to be assessed and then the precise span can be more accurately calculated.

Taking the above formula and assuming the following: Ledger = 200×100 mm. $\sigma = 7.4\,\text{N/mm}^2$, f = 5600, s = 1500 (joist span). Then by substitution,

$$L = \sqrt{\frac{8 \times 10^6 \times 7.4 \times 100 \times 200^2}{6 \times 5600 \times 1500}} = 2167\,\text{mm}$$

say 2000 mm.

With one joist central on the span the layout will now be as in Fig. 86 f_i = load from each joist which is the same as the load carried by the joist

$$= \frac{5600 \times 600 \times 1500}{10^6} = 5040\,\text{N}.$$

If now symbols are substituted for figures as in Fig. 87 a formula can be evolved for the bending moment using the standard method.

Thus $R_L = R_r = 1.5 f_i$. Taking moments to the right about the centre of the beam

FORMWORK

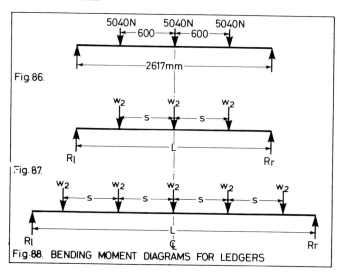

Fig. 86.

Fig. 87.

Fig. 88. BENDING MOMENT DIAGRAMS FOR LEDGERS

$$ACWM = \frac{L}{2}1.5 f_i = 0.75 Lw_2$$

$$CWM = s_i f_i.$$

Therefore maximum $BM = 0.75 Lw_2 - s_i f_i = f_i(0.75L - s_i)$ and the equation is

$$f_i(0.75L - s_i) = \frac{\sigma bd^2}{6}.$$

The evolution now is progressively:—

$$0.75L - s_i = \frac{\sigma bh^2}{6f_i}$$

and

$$L = \frac{1}{0.75}\left(\frac{\sigma bh^2}{6f_i} + s_i\right).$$

Substituting numerical values

$$L = \frac{1}{0.75}\left(\frac{7.4 \times 100 \times 200^2}{6 \times 5040} + 600\right) = 2105 \text{ mm}.$$

The percentage error of treating the load on the ledger as UDL is therefore

$$\frac{2167 - 2105}{2105} \times 100 = 2.9\%.$$

In view of the normally high factor of safety taken with timber construction and of other unknown estimated factors, the method of calculation for ledgers assuming a UDL is often considered good enough.

To check for deflection, erring a little on the safe side, the precise BM may be divided by L/8 to give a load value which may be regarded as a UDL. This may then be used in the formula

$$D = \frac{5}{384}\frac{FL^3}{EI}$$

to give the deflection.

$$F = \frac{8f_i(0.75L - s_i)}{L}.$$

$$= \frac{8 \times 5040 (0.75 \times 2105 - 600)}{2105} = 18747 \text{ N}.$$

Then

$$D = \frac{3}{384}\frac{18747 \times 2105^3 \times 12}{7040 \times 100 \times 2000^3} = 4.85$$

ratio deflection to span

$$= \frac{4.85}{2105} = \frac{1}{453}$$

which is satisfactory.

If one joist is assumed to be in mid span then an odd number of joists will always be presumed to be carried. A larger and longer ledger might therefore carry 5 joists (Fig. 88).

The same principle may be used to work out the bending moment.

Thus $R_L = Rr = 2.5f_i$. Thus taking moments about mid span to the right

$$ACWM = 2.5f_i 0.5L$$

$$1.25 f_iL$$

$$CWM = f_is_i + f_i2s_i$$

$$= 3f_is_i.$$

Maximum BM therefore $= f_i(1.25L - 3s_i)$.

Loading on props

It is reasonable to assume that the loading of the props can be calculated as from a UDL. The area of concrete to be carried may then be taken as ledger span × joist span (Fig. 85) which in the previous example will be 2105 × 1500 mm. Therefore total load carried by each prop will be

$$\frac{5600 \times 2105 \times 1500}{106} = 17682 \text{ N}.$$

In present day construction proprietary equipment is generally used to support slab forms.

From the manufacturers tables for 'Acrowprops' their 2 × size, extended from 6 ft. 6 in. to 11 ft. (given in imperial measure) have a fail load of 17472 lb. which in SI is 77667 N.

The safety factor therefore $= \frac{77667}{17682} = 4$ which is twice the minimum needed.

Design of vertical forms

As stated already, the hydrostatic pressure in vertical forms from the wet concrete depends mainly upon the head of unset concrete, but also to some extent upon other factors involved in pouring and placing.

The following data is based on investigation carried out by the Construction Industry Research and Information Association:

In computing the maximum pressure of concrete in vertical forms the following factors have to be considered:

1. Density of concrete $\Delta(kg/m^3)$
2. Workability of the mix (slump in mm)
3. Rate of placing $R(m/h)$
4. Method of concrete discharge (possible impact load)
5. Concrete temperature T (deg. Celcius)
6. Percentage of continuity of vibration
7. Height of lift (H in mm)
8. Least dimension of the section of the wall or column

22

9. Reinforcement detail
10. Stiffness of the formwork structure

For convenience, normal conditions of vibration, condition of formwork and amount of reinforcement is assumed.

As the pressure is likely to be very much more than will occur in horizontal slabs this is given in kN ($= 1000$ N) per metre square. Other symbols have different unit values and must be expressed as such for the formulae to be effective. These formulae are:

1. Where the least dimension of the wall or column is more than 500 mm and the maximum pressure is limited by the stiffening of the concrete before the end of the pour.

$$P_{max} = \left(\frac{\Delta RK + 5}{100}\right) kN/m^2.$$

2. Where the least dimension 'd' of the column or wall is less than 500 mm there will be some reduction in hydrostatic pressure due to the aggregate constituent of the concrete arching between the forms and so not adding its weight to the fluid. Then

$$P_{max} = \left(3R + \frac{d}{10} + 15\right) kN/m^2.$$

3. When the rate of fill, temperature, etc., are such that the whole of the lift is poured before any concrete sets. The max. unit pressure then is merely due to the density multiplied by the height of the lift. Then

$$P_{max} = \frac{\Delta H}{100} kN/m^2.$$

The denominator of 100 merely changes the kg value of the density to kN in the answer.

If the concrete is to fall freely more than 2 m in deposit an additional value of 10 kN/m² should be allowed for impact for the whole height. The constant K in formula 1 is a correction factor to allow for concrete temperature and workability derived from the table below.

Workability mean slump mm	Concrete temperature °Celsius					
	5	10	15	20	25	30
25	K = 1.45	1.10	0.80	0.60	0.45	0.35
50	K = 1.90	1.45	1.10	0.80	0.60	0.45
75	K = 2.35	1.80	1.35	1.00	0.75	0.55
100	K = 2.75	2.10	1.60	1.15	0.90	0.65

It should be noted that formulae 1 and 2 both give values based on the rate of fill irrespective as to the height of the lift. Thus under conditions of very fast fill at say low temperature the calculated pressure which depends upon the depth of unset concrete in the box might be equivalent to that which would occur at a depth say 5 m from the top but if the fill was only 3 m deep this would never be reached.

Thus the maximum pressure which can occur apart from impact load already mentioned is given by formula 3 which should be used in preference to the other when they are likely to give values too high. To find the depth in m from the top of the lift at which maximum pressure occurs divide P_{max} by Δ in kN, i.e. $\frac{\Delta}{100}$.

If in formula 1 or 2 this is greater than the height of the lift values are too great and formula 1 should be used.

Examples of calculation of maximum pressure

EXAMPLE 1: Given – Height of lift $= 4$ m, $\Delta = 2400$ kg/m², $R = 4$ m/h, $t = 15°$ c. Slump $= 50$ mm. From the tables of workability and temperature $K = 1.10$. Using formula 1.

$$P_{max} = \frac{2400 \times 4 \times 1.10}{100} = 105.6 \ kN/m^2.$$

Given H_i is the depth from the top of the lift of the level of maximum pressure and

$$\Delta kN = \frac{2400}{100} = 24 \ kN \text{ then } H_i = \frac{105.6}{24} = 4.4 \ m.$$

But the height of the lift is only 4 m so the maximum pressure must be obtained by means of formula 3 so

$$P_{max} = \frac{2400 \times 4}{100} = 96 \ kN/m^2.$$

EXAMPLE 2: Given wall $= 250$ mm thick height of lift $= 5$ m and rate of pour $= 3$ m/h. Then from formula 2

$$P_{max} = 3 \times 3 + \frac{250}{10} + 15 = 49 \ kN/m^2.$$

Then depth of level of maximum pressure $= 49/24 = 2.041$ m.

Design of formwork for wall given the maximum pressure in the forms

Wall forms may be made with horizontal sheeting, vertical studs and horizontal walings or vertical sheeting, horizontal studs and vertical soldiers.

In the first case the capacity of the sheeting to take the maximum hydrostatic pressure, governs the spacing of the vertical studs. The strength of the vertical studs governs their various spans or the spacing of the walings whilst the strength of the walings governs the positions of the tie bolts through the concrete.

In the alternative construction, the horizontal studs are spaced closer together downwards to the level of maximum pressure and below to the maximum safe capacity of the vertical outer grained plywood. Vertical soldiers are spaced to suit the safe span of the horizontal studs at maximum pressure, while the tie bolts through the concrete are placed closer together at the bottom to the capacity of the soldiers to take the increasing hydrostatic pressure.

Examples of wall form design using horizontal sheeting, vertical studs and horizontal walings

Assume height of wall $= 4$ m and maximum pressure on forms $= 50$ kN/m². Then $H_i = 50/24 = 2.083$ m say 2 m. Assuming that concrete will be dropped 2 or more metres an impact load of 10 kN must be added for the full height of the lift. This is indicated by the pressure envelope shown in the graph in Fig. 89.

Using vertical studs and 19 mm plywood sheeting the criterion for the spacing of the studs will be the bending strength of the plywood (deflection is seldom necessary to be considered where heavy loading is involved). So

FORMWORK

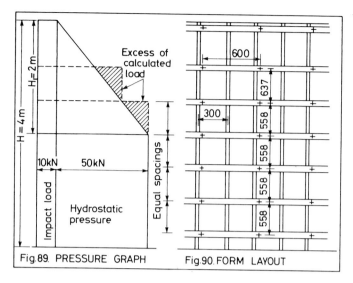

Fig. 89. PRESSURE GRAPH   Fig. 90. FORM LAYOUT

$$\frac{FL}{10} = \sigma Z. \qquad \frac{fsL^2}{10 \times 10^6} = \sigma Z \quad so \quad L = \frac{10^7 \sigma Z}{fs}.$$

From tables with adjustments bending $\sigma = 7.4\,N/mm^2$. Z for 19 mm unsanded 7 ply $= 4.9 \times 10^4$ per m. wide. f expressed in N/mm^2 (to agree with σ) $= 6 \times 10^4$

By substitution

$$L = \sqrt{\frac{10^7 \times 12.2 \times 4.9 \times 10^4}{6 \times 10^4 \times 10^3}} = 315\,mm,$$

as the studs are 50 mm thick if they were spaced at 330 mm centres this would still give 35 mm less than the calculated value in the clear.

Vertical studs

Using 100×50 mm studs these will be supported by walings at spacings which may be increased above the line of maximum pressure. In calculating the spacings of walings the formula is now

$$L = \frac{10^7 \sigma bh^2}{6fs} \quad \text{(incorporating the full section modulus).}$$

As the equation is to be used several times with only w_i changed this can be left in as the denominator of a simple coefficient. Thus

$$L^2 = \frac{10^7 \times 7.4 \times 50 \times 100^2}{6 \times 330\,f}$$

$$L^2 = \sqrt{\frac{1.8686 \times 10^{10}}{f}}$$

Below the line of maximum pressure walings will be equally spaced at

$$L = \sqrt{\frac{1.8686 \times 10^{10}}{6 \times 10^4}} = 558\,mm.$$

Above this line the same spacing will still give a reasonable approximation where the pressure may be reduced as in the graph (Fig. 89) where the pressure will be 4.6×10^4 N/m^2.

The next spacing will now be

$$\sqrt{\frac{1.8686 \times 10^{10}}{4.6 \times 10^4}} = 637\,mm.$$

From the graph it is obvious that no further supports are needed below the top of the form.

Finally the ties in the walings are usually placed against a stud for practical reasons. Thus the load from the stud is taken entirely by the tie. Assuming a 4 ft. (1.2 m) wide sheet of ply with 4 stud spacings ties will be spaced at 600 mm centres, and the waling will only support one central stud.

Therefore giving spacing of studs at 300 mm crs and spacing of walings at 558 mm crs then

$$f_i = \frac{300 \times 558 \times 6 \times 10^4}{10^6}$$

$f_i = 10^4$ N. BM for central loading on a partly continuous beam is

$$\frac{6FL}{32}$$

$$\frac{6FL}{32} = \frac{\sigma bh^2}{6}$$

$$L = \frac{320bh^2}{6 \times 6F}.$$

Assuming walings are double 100×50 mm and other values as before

$$L = \frac{32 \times 7.4 \times 100 \times 100^2}{6 \times 6 \times 10^4} = 657\,mm$$

so that ties at every other stud will be satisfactory. Fig. 90 shows the arrangement of the timbers as calculated.

It should be noted that these examples serve only to introduce the basic principles of design calculations. There are various textbooks available for further study.

CHAPTER 3

Timber Floor Construction

GROUND FLOORS

In no other part of a building is the health and durability of timber so much at risk as it is in timber ground floors, which are most of the time the starting point for the attack of one of the two common fungi known as 'dry rot' and 'wet rot'. Yet in spite of this, wooden floors are almost cetainly still the most popular, primarily for the reasons that they are economically constructed, are warm to the feet, provide an easy fixing for carpeting and other flooring materials or when constructed of a suitable decorative timber, extremely attractive in appearance.

However, if one excludes the effects of national disasters such as floods, it will be found in most cases that the unacceptable presence of moisture in buildings, leading to timber decay is invariably due to bad design or imperfect construction. If therefore the rules of sound construction as indicated by the Building Regulations are followed little trouble is likely to arise.

In general terms to prevent decay in timber in building, due to excess moisture, there are two requirements:

1. The timber must not be allowed to rest in contact with any other porous material (e.g. brickwork or concrete) which is itself in contact with the ground or exposed to the weather.
2. It must not be continuously exposed to humid conditions of air likely to create dampness by condensation.

In ground floor construction these requirements are met:

1. by providing a damp proof course under wall plates and against all vertical surfaces with which any timber is in contact; and
2. by ensuring that there are sufficient air bricks or other means of ventilation giving a chance for a reasonable movement of air around all timbers under the floor.

Fig. 91, which is a section through the external wall and a hollow timber floor illustrates the Building Regulations governing this construction. Insofar as they apply to the site carpenter these are briefly as follows:

There must be an oversite layer of concrete which must be either 100 mm thick over a layer of hardcore or it can be 50 mm thick but must then be laid on a dpm at least 1000 gauge with sealed joints, bedded on smooth material such as sand to avoid damage to the sheeting, the concrete must be to a minimum specified mix. The surface of the concrete must be not lower than the highest level of the adjoining ground or the surface must be drained.

There must be a clear space free from debris of at least 75 mm to the underside of any wall plate and at least 125 mm to the underside of the floor joists. This space must be sufficiently ventilated to ensure a movement of air without stale pockets. There must be damp courses to prevent the direct entry of moisture.

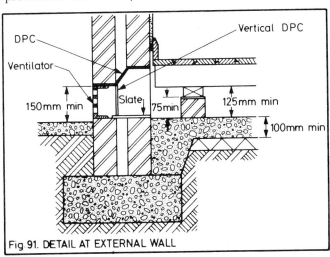

Fig. 91. DETAIL AT EXTERNAL WALL

It will be noticed that the air brick is sealed from the wall cavity with a flexible damp course over the top.

Although the detail as sketched does conform to the regulations it would be an advantage to have floor and damp course one brick higher so as to lift the air brick or ventilator higher above the external paving. This paving should of course have a definite fall away from the house wall. The ends of the joists are shown carried on an isolated sleeper wall rather than built-in or carried on a brick offset. This makes for a better circulation of air around the joist ends which are vulnerable. All sleeper walls should be built open or 'honey-combed' as shown in Fig. 99.

TIMBER FLOOR CONSTRUCTION

Fireplaces

In trimming the floor around the fireplace the fire hazard also has to be considered. The Building Regulations cover this in ruling that there shall be an incombustible hearth extending 150 mm on either side of the opening and at least 500 mm in front of the jambs.

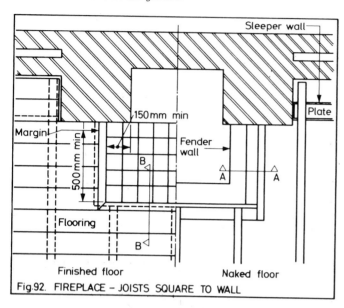

Fig.92. FIREPLACE – JOISTS SQUARE TO WALL

The requirements, as far as the carpenter is concerned, are to support the ends of the floor joists and the edges of the flooring around the opening and to contain the concrete and the tile hearth on the hardcore within the fender walls provided by the bricklayer. This is shown in plan in Fig. 92 and in section A-A Fig. 93 and section B-B Fig. 94. Note that the concrete hearth and fender wall are both isolated by damp proof courses.

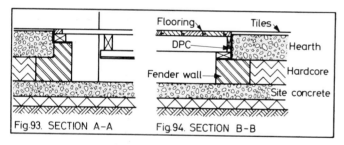

Fig.93. SECTION A-A   Fig.94. SECTION B-B

The floor boards are finished against the hearth with a margin mitred at the front corners. This may be softwood but is sometimes hardwood even in a softwood floor. A better job, seldom done, is to rebate the edges of the floor boards and pin and glue half-thickness margins into the rebates as in Fig. 95. Another alternative is to fit a margin to the board ends only and mitre this into the solid flooring at its side, as in Fig. 96.

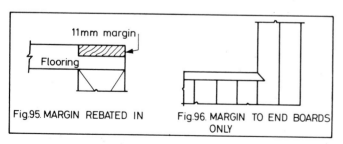

Fig.95. MARGIN REBATED IN   Fig.96. MARGIN TO END BOARDS ONLY

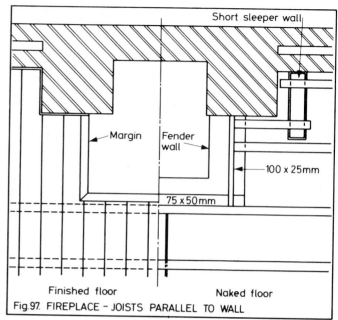

Fig.97. FIREPLACE – JOISTS PARALLEL TO WALL

The method used to carry the ends of the joists when they are deprived of their normal support sometimes requires some little ingenuity and, in Fig. 97, the first joist against the wall and a short one against the breast are carried by a short sleeper wall especially built. As an alternative, the end of the first joist could have been built into the breast with a short trimmer cut between that and the next to take the end of the other short joist. In this case the built-in end would have had to be treated with preservative. This is still not a good job as the breast is part of the outside wall.

Floor joists

When laying out the floor joists for a room the aim should be to use the minimum amount of timber by keeping to maximum allowances for spacing of joists, etc. Fig. 98 represents a typical plan of an L-shaped room of the combined lounge-diner type, with possible joist layout.

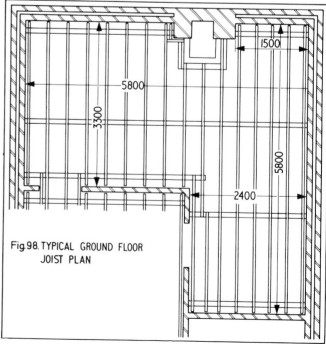

Fig.98. TYPICAL GROUND FLOOR JOIST PLAN

Spacings of joists in a given area seldom work out at an exact multiple of the given maximum allowance so that generally the positions of individual joists or groups of joists can be adjusted for the greater convenience without anywhere exceeding the permissible maximum. In the writer's opinion these details can be better worked out to scale from a simple single line diagram with a few calculations than by moving long timbers about the floor.

The exact size of floor joists to meet a special condition can be calculated but for most ordinary circumstances values to meet known conditions of loading can be taken from the building regulations. These give the choice of sizes related to span or visa versa for strength classes SC3 (table B3 and SC4 (table 4) the grades of the various species which relate them with these classes being given in another table. From table B3 with a dead load not exceeding $0.5 \, kN/m^2$ joists size $47 \times 100 \, mm$ and spaced at $400 \, mm$ can span $2 \, m$.

From Fig. 98 the width of the room $= 5800 \, mm$, and distance of centre of first and last joist from the wall $75 \, mm$, then distance between first and last joist $= 5800 - 150 = 5650 \, mm$.

Then the number of joist spacings

$$= \frac{5650}{400} = 14 + 1$$

for a remainder $= 15$ spacings. 15 spacings of 400 (max). $= 6000 \, mm$, so we have $6000 - 5650 = 350 \, mm$ to juggle with

Starting from the right, if one joist is placed for convenience 50 mm from the chimney breast, spacing of joists between

$$= \frac{1500 - 150}{4} = 338 \, mm \, crs.$$

The width covered by the 11 remainder spaces $= 5650 - (1500 - 150) = 4300 \, mm$. Therefore remainder of joists may be equally spaced at

$$\frac{4300}{11} = 390 \, mm \, crs.$$

which is just within the maximum limit.

The joist to the left of the chimney could therefore be moved in $7 \times 9 = 63 \, mm$ (by increasing spacings to maximum) which would not be sufficient to avoid trimming shown, although in practical terms the error is so slight it would be ignored.

Assuming end sleeper walls are 150 mm from the main walls the centre line distance between them $= 5800 - 400 = 5400 \, mm$ so the spacings will be

$$\frac{5400}{3} = 1800 \, mm.$$

These also could be adjusted up to the maximum spacing of 2000 mm if it were required. Although the end sleeper walls are shown 150 mm in the clear, this distance could be easily increased to 300 mm if there were any advantage in so doing by reducing the intermediate spans.

Once these details have been worked out spacings can be marked on a pair of battens from which the joists can be spaced and temporarily nailed.

Ventilation

As already stated, underfloor ventilation is very essential to ground floor construction and it is often necessary to advise the occupier not to cover the air bricks accidentally with garden soil, or deliberately to prevent draughts through the floor joints or under the skirting as has been seen by the writer. Air bricks should be opposite each other to ensure a constant flow of air and should be spaced not more than 1.8 m apart. Building regulations require a minimum of $3000 \, mm^2$ of nett ventilation space perm metre run of wall.

It is however advisable to increase this value under damp conditions, where the minimum depth of underfloor space is provided, or where there is a wide floor area as in bungalows. Ventilation openings should be left in partition walls and any possible stagnant corners as may occur against chimney breasts relieved. One problem occurs when the adjoining room has a solid floor. This may be overcome by providing ducts under the floor to the outer air (Fig. 99) to conform with regulations any pipes needed to carry ventilation air must be at least 100 mm diameter.

A watch should be kept on damp-courses which occur at different levels as in Fig. 99 where a short vertical membrane connects the damp-course in the wall to that under the solid floor. Without this the otherwise protected partition wall would be exposed to rising damp from the oversite concrete.

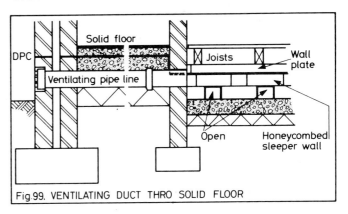

Fig.99. VENTILATING DUCT THRO SOLID FLOOR

The greatest danger of decay in a new building undoubtedly arises in the first 12 months after completion, due to the presence in the brickwork, plaster, concrete, etc., of the 'moisture of construction'. The condition may be aggravated by hurried completion; or timbers left unprotected and used in a wet condition. Where, due to site conditions, trouble is anticipated, it may be advisable to treat joists, backs of skirtings and undersides of floorboards with an inodorous wood preservative, such as Cuprinol, merely brushed or sprayed on. This will protect the timber through the dangerous drying out period.

If decay is discovered it is most likely to be wet rot (Coniophoro cerebella), the action of which will be stopped immediately by drying, although of course any weakness left is there for good. A very careful investigation should be made before the drastic treatment associated with dry rot (Merulius lacrymans) is given.

The moisture content of ground floor joists should not exceed 18 percent, and if they are drier than this, swelling in situ is not likely to give trouble. Normal softwood flooring need not be kiln dried. Some shrinkage may ultimately occur but this will not be important if the floor is to be covered with carpet or vinyl. The moisture content should however be well below 20 percent to avoid risk of decay.

Softwood and hardwood strip flooring should however be kiln dried to the percentage MC at which it could be

TIMBER FLOOR CONSTRUCTION

expected to stabilise under ultimate conditions of occupancy. The recommended values are 14 percent where there is to be intermittent heat and 12 percent MC for continuous heating conditions. Where there is to be underfloor heating the MC should be reduced to 7 percent.

It is essential that kiln dried flooring should be prevented from absorbing moisture at any time after being brought on the site. To this end it should not be delivered until needed, when the genral site conditions should be as follows:

1. Windows should be glazed and external doors hung.
2. Plastering should be completed.
3. Any screeds to solid floors should be perfectly dry.
4. Heating should have been put on and the building dried out.

The heating should be continued until the building is in occupation.

Softwood floors

Heading joints between ends of boards, usually butted, should be staggered as much as possible, and should come midway on the joists. Traps for access should be screwed into position and the bottom face of the grooved side removed to form a rebate for ease of removal. Boards up to 100 mm wide on face should be fixed through the face with one nail to each joist. Boards over this width should have two nails to each joist, about 12 mm from the tongued edge and 15 mm from the grooved edge. Nails may be cut nails or lost heads. They should be $2\frac{1}{2}$ times the thickness of floorboards in length and punched below the surface.

Cramping floorboards

All flooring must be cramped up tight before top nailing. This can be done in a variety of ways. The most efficient method is by means of flooring cramps. These grip the joist by means of a levered cam action, enabling a screw operated head to thrust against a protective batten placed against the last board and so tighten the joints. If the boards are laid tongue out the batten should be grooved to fit and is best cut from an odd piece of flooring. It is sometimes better not to drive the last row of nails each time, thus enabling the next tongue to be more easily inserted.

An alternative but less efficient method is known as 'folding'. Five or more boards according to width are laid and pushed hand tight. The back edge of the last one being marked on the joists. The last board is then moved forward 6 to 12 mm according to the apparent hand tightness and securely nailed. The other boards are then arched into position as shown in Fig. 100. A couple of short boards are then laid across the joint lines and jumped on at A.

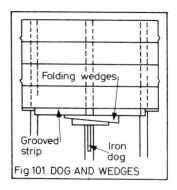

Fig.101. DOG AND WEDGES

An alternative method using iron dogs (driven into the tops of the joists) from which pressure is exerted by means of folding wedges is shown in Fig. 101. It will be found convenient if the inner one is kept long, a hammer held against the short one and the other driven against it.

The last few boards cannot be cramped by any of the above methods, but they can be tightened by levering off the wall as shown in Fig. 103 a block being driven down behind the lever to keep it tight while nailing.

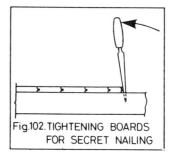

Fig.102. TIGHTENING BOARDS FOR SECRET NAILING

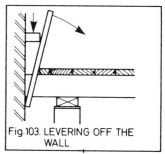

Fig.103. LEVERING OFF THE WALL

Strip flooring

When the floor itself forms a decorative feature it is usually constructed from hardwood and secretly nailed. Each board may be nailed through the tongue as in Fig. 104 and punched clear, each tongue then holding the grooved edge of the next board. To reduce the amount of shrinkage on each board, the widths are made narrow, from 70 to 100 mm, hence the name strip flooring. The heading joints are also often tongued and grooved when they do not need nailing, or fixing on a joist. Fig. 105 shows two alternative sections. The splayed tongue makes for easy and more solid nailing while the hollow back on the other section is designed to give a positive bearing on the joist in the event of slight shrinkage distortion.

It will be appreciated that the boards have to be fixed one at a time so cramping is not a practical proposition and the boards are usually levered tight by means of a stout chisel driven into the top of the joist as shown in Fig. 102.

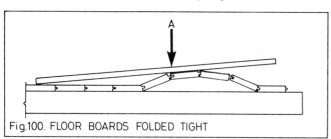

Fig.100. FLOOR BOARDS FOLDED TIGHT

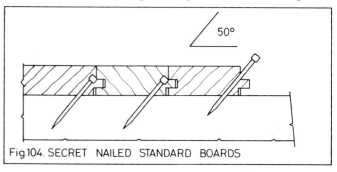

Fig.104. SECRET NAILED STANDARD BOARDS

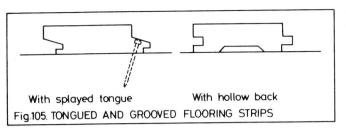

With splayed tongue | With hollow back
Fig.105. TONGUED AND GROOVED FLOORING STRIPS

Timber boarded solid floors

The boards are nailed to battens or fillets which may either be secured to a cement sand screed by means of galvanized iron clips or imbedded in the screed. In both cases, the Building Regulations specify a subfoundation of a 100 mm layer of concrete on hardcore on which is laid a damp-proof membrane of asphaltic bitumen or coal tar pitch at least 4 mm thick.

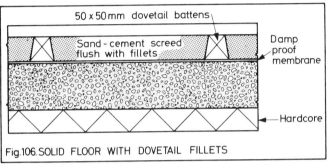

Fig.106. SOLID FLOOR WITH DOVETAIL FILLETS

Using imbedded dovetail fillets as shown in Fig. 106, these are treated with preservative in accordance with provisions of BS 3452 or 4072 and the 1:3 sand-cement screed flushed up between them. This should be allowed to dry thoroughly before laying the floorboards. The undersides of the floorboards should preferably be treated with a clear preservative or the boards laid on a water-proof felt.

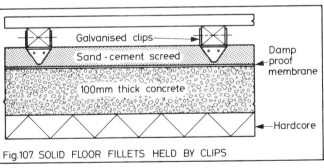

Fig.107. SOLID FLOOR FILLETS HELD BY CLIPS

In Fig. 107 a cement-sand screed at least 38 mm thick is laid on the dpc and in this are imbedded the galvanised floor clips at from 300 to 600 centres. (The stouter the fillets, the wider the clip spacing.) The fillets are spaced at about 400 mm centres. The fillets and the undersides of the floor-boards are treated with preservatives respectively as before.

Wood block floors

These are generally laid by specialists. The preparation requires, as for boarded solid floors, a damp-proof membrane under the screed although if the blocks are laid with a continuous hot adhesive on the screed, the Building

Regulations recognise this as the damp-proof membrane. The bottoms of the blocks are dovetailed to form a key into the adhesive while individual blocks key to each other by means of tongued and grooved or dowelled joints.

An expansion joint is necessary around the walls. This should be filled with compressive material to allow swelling of floor but prevent creep, and not left open.

Parquet floor finishes

These consist of special thin wood blocks of decorative hardwood 6 to 9 mm thick laid to various patterns and glued and pinned to a suitable sub-floor.

The best base giving a suitable flat surface for fixing is plywood (external grade WBP).

If there is an existing timber floor this should be levelled off, all nails punched in and then surfaced with 3 mm plywood fixed with special clout nails with the grain of the outer ply diagonal to that of the wood floor. The parquet blocks are then glued and pinned to the plywood.

On a new floor large sheets of 15 mm plywood are nailed direct to the joists, joints being supported by noggings where necessary. Joints should be left slightly open and filled with mastic. Nails should be driven at about 300 mm centres not nearer to the edges than 6 mm. If there are any irregularities such as variation in ply thicknesses, these should be sanded off.

Fixing of skirtings

Skirtings should be mitred together at external (and internal) angles, and butt jointed (tongued and grooved on special jobs) on the flat with mouldings scribed. A mitre is first cut on the moulding to give the outline for the scribe which can generally be cut with a coping saw. Fig. 108 shows a mitre box suitable for the purpose. The part of the

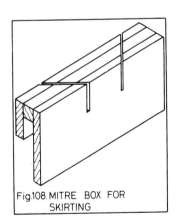

Fig.108. MITRE BOX FOR SKIRTING

skirting which runs through on the internal angle should preferably be let into the plaster. On a good job the skirting should be scribed to the floor. In alterations to old buildings, particularly when two or more rooms are converted into one, floors may be far from level and it is as well before starting to fix the skirting to check the floor with a straight-edge to get an idea of the depth of scribe required to accommodate all irregularities.

TIMBER FLOOR CONSTRUCTION
UPPER FLOORS

The construction of upper floors differs from that of ground floors in that the floor as well as a ceiling have to be carried without intermediate support for the full span of the room.

The risk of decay through damp is lower. Indeed if the building is soundly constructed with clean cavities in the walls, little trouble need be anticipated in this direction.

As the span between walls increases, the joists have to be correspondingly deeper; and over a span of more than 4.9 m (16 ft), simple construction is no longer viable and more complex design becomes necessary. According to their minimum spans and relative constructions, floors with timber joists may be classified as follows:

SINGLE FLOORS (Fig. 109): For spans up to 4.9 m where joists span from wall to wall.

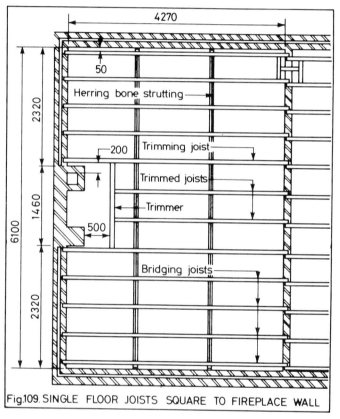

Fig.109. SINGLE FLOOR JOISTS SQUARE TO FIREPLACE WALL

DOUBLE FLOORS (Fig. 111): For spans up to 5.5 m (18 ft). The length of the room is divided into bays 2 to

3 m long by transverse beams (usually BSBs) which carry the timber joists. In relation to this construction these beams are called 'binders'.

FRAMED FLOORS (Fig. 112): For spans over 5.5 m heavy transverse BSBs, termed girders span the shortest way of the room in bays of from 3 m to 3.5 m long. These in turn carry the binders, themselves spaced at from 2 to 3 m which take the timber joists.

The layout of floor joists in domestic work is commonly governed by the positions of fireplaces, as the Building Regulations prohibit the placing of timbers in proximity to hearths and flues.

Those regulations which particularly affect joist positions say that no timber shall be placed closer than 500 mm to the face of the chimney breast nor nearer than 200 mm to the flue or fireplace opening so that no structural timber can be supported by the breast.

Fig. 109 illustrates the layout of the floor when the joists are at right angles to the wall containing the fireplace. Two heavier joists termed 'trimming joists' are taken into the wall on either side of the breast. As one side of the breast carries a flue the joist that side must be kept at least 200 mm from the flue lining. On the other side, the joist may be taken close.

All joists should have a bearing of at least 75 mm on the wall, or other support, at their ends. A short joist of the same section and known as a 'trimmer' is cut between them. This in turn takes the ends of the shorter 'trimmed joists'.

Fig 110 shows the arrangement when floor joists are parallel to fireplace.

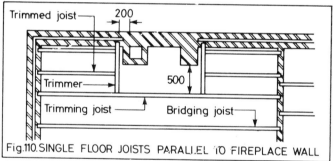

Fig.110. SINGLE FLOOR JOISTS PARALLEL TO FIREPLACE WALL

In Fig. 109 an example has been taken of a room 6100 mm long and 4270 mm wide.

In relation to timber sizes and layout, the Building regulations table B3 and B4 give inter-related joist spans sizes

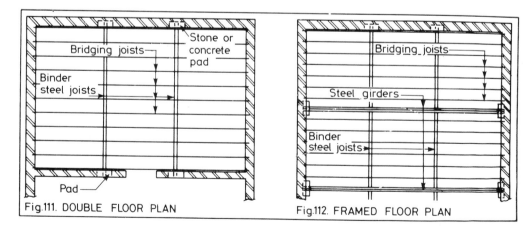

Fig.111. DOUBLE FLOOR PLAN

Fig.112. FRAMED FLOOR PLAN

and spacings for SC3 and SC4 strength classes of timber. The tables can only be used for single occupied domestic buildings not more than three storeys in height for which imposed unit loadings of 1.44 N/m^2 are assumed. In other conditions the design sizes or spacings and spans will have to agree with BS 5268 part 2.

The positions of the trimming joists condition the spacing for both trimmed and bridging joists involving one or two extra members over what would otherwise be required for a straight layout. If the actual spacings are calculated beforehand, some of the extra expense may be recovered by using smaller joists within the framework of the Building Regulations tables.

In Fig. 109, as there is a central fireplace 1460 mm wide, there are spaces either side of 2320 mm. Assuming, to start with, that 21 mm floor boards could be used.

Maximum permitted spacing of joists = 635 mm centres.

Width overall of trimming joist centres = 1450 + 50 + 75 mm = 1575 mm.

Then number of intermediate spaces

$$= \frac{1575}{635} + 1$$

for a remainder = 3.

Therefore spacing of trimmed joists

$$= \frac{1575}{3} = 525 \text{ mm}.$$

Similarly number of spaces between bridging joists = 4, so spacing of bridging joists

$$= \frac{2195}{4} = 548.75 \text{ mm}.$$

Span of bridging joists = say 4300 mm. Span of trimmed joists = 3350 mm.

Maximum span of floor boards = 548 mm.

Flooring to span 600 mm need only be 19 mm thick.

Bridging joists spaced at 600 mm, 4.70 m span require to be 63 mm by 225 mm section.

Trimmed joists spaced at 600 mm, 3.70 span need only be 38 mm by 225 mm section.

The general rule for trimmers and trimming joists is that they should be 3 mm thicker for every trimmed joist carried so that they could be 75 mm by 225 mm, being the nearest commercial size over.

To prevent lateral movement or buckling of long joists particularly as they are deeper in relation to their thicknesses they need strutting at regular intervals of span. Some authorities give a general rule that strutting should be spaced at distances of 50 joist thickness which would mean one row of struts only in the example given. In my opinion, however, this cross bracing does considerably stiffen the floor and closer spacing will therefore reduce the risk of cracks in a plaster ceiling.

Alternative types of strutting are solid, and herring-bone strutting. The latter is superior in that whereas shrinkage in the thickness of the joists will cause solid strutting to loosen the pro rata greater shrinkage in depth which accompanies this will cause the 'scissors' struts to close and tighten.

Fig. 113 is a pictorial sketch showing the herring-bone strutting in position between the joists and indicates the method of marking out and cutting. To mark out, two parallel lines are struck across the joists in the position

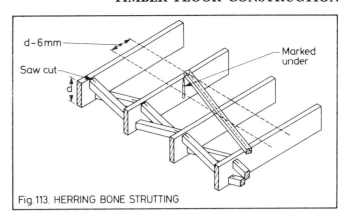

Fig. 113. HERRING BONE STRUTTING

where the strutting has to go, the distance between the lines being the joist depth minus 12 mm. The batten is then placed across the lines diagonally as shown and marked under against the joist. Thus the struts may be cut to fit each individual space.

A kerf from a tenon saw in the splayed end as shown, enables the nail to be driven without splitting the timber. With permitted low domestic ceilings, an apprentice working off a plank on stools can give some help in this from below. It is essential that the combined thrust of the struts should be countered at each end. This is done with folding wedges against the wall (Fig. 109).

Forming joints on the trimmer

There are a number of different methods of forming the joint between the trimmed joist and the trimmer and the trimmer and the trimming joist. In each case the supporting timber must be weakened as little as possible.

The best and most scientific joint, although its labour cost has tended to make it obsolete is the tusk tenon. It is illustrated by a sketch in Fig. 114 and in sectional elevation in Fig. 115. It is designed on the scientific fact that the greatest stresses in a beam are in the extreme fibres. If it has to be cut away this should be progressively more towards the centre of the depth.

Two dotted diagonal lines in the section in Fig. 115 forming between them two vertical triangles represent the amount of bending stress at any depth in the beam. Thus,

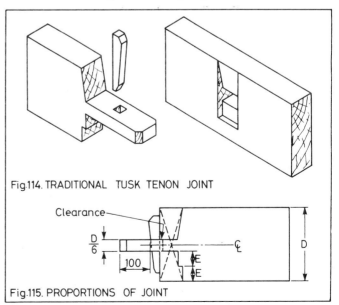

Fig. 114. TRADITIONAL TUSK TENON JOINT

Fig. 115. PROPORTIONS OF JOINT

ignoring shear, the top and bottom fibres are fully stressed; a quarter way down they are only half stressed; while in the centre there is no bending stress at all.

To mark out, draw the section of the joist and the diagonals, put the through tenon in the centre at 1/6 depth and make tusk below this one half the remaining depth. The tusk should just touch the diagonal as shown.

The mortise in the tenon to take the wedge or key should be chopped back so that it is clear of the back of the key when tightening. The joint should be accurately made so that when assembled both the underside of the tenon and the tusk take a firm bearing.

A common fallacy is that only the area within the dotted lines is under stresses in bending and that the timber outside of these lines may be cut away with impunity. This is not so. The whole width of section at every level is under stress although reducing towards the neutral axis, and if substance is removed anywhere there will be some loss of strength. If however the cutting is kept without the dotted lines, the loss of strength will be minimal.

Another fallacy is that the tusk tenon should be kept above the neutral axis in the compression area; the idea being that the tenon fits into the mortise and replaces the compression timber taken out when chopping the mortise. If the tenon had to be so tight to give this effect, it could not be got into the mortise.

Figs. 116 to 121 show alternative methods of connecting trimmed timbers and trimmers. These are sufficiently strong for most purposes, but when there is a loss of strength due to the substance being cut away the overall thickness should be increased to allow for this.

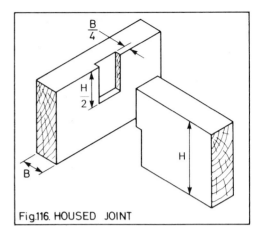
Fig.116. HOUSED JOINT

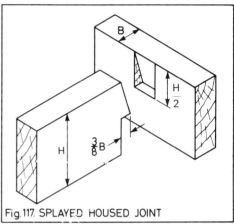
Fig.117. SPLAYED HOUSED JOINT

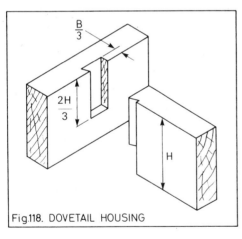
Fig.118. DOVETAIL HOUSING

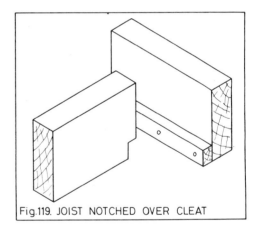
Fig.119. JOIST NOTCHED OVER CLEAT

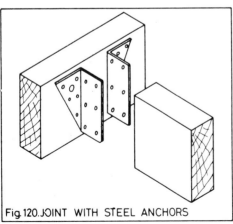
Fig.120. JOINT WITH STEEL ANCHORS

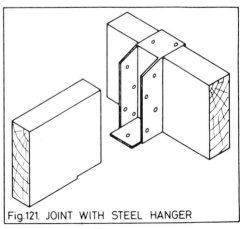
Fig.121. JOINT WITH STEEL HANGER

Fig. 116 shows a housed joint. The cut away should not exceed the proportions shown. In the splayed housing in Fig. 117, there is very little loss of strength, but the joint needs to be well nailed. The dovetail housing Fig. 118 is useful where a short projection is used to form an L-shaped opening in the floor, say for a stairway. If this is taken too near the centre of the span of the supporting member, the timber thickness should be increased.

Fig. 119 shows a joist notched over a cleat. The bearing capacity of the beam depends upon the security of the nails. Figs. 120 and 121 show methods of supporting trimmed timbers using proprietary anchors and hangers. The joint is made using 31 mm galvanised nails with serrated shanks. In Fig. 120 the strength depends entirely upon the nails.

Supporting the floor joists

The commonest way of supporting floor joists at the ends is by building them into the wall, the joist having a bearing of at least 75 mm at the ends. A better method is to bed them on a wrought iron plate as in Fig. 122. Fig. 123 shows the joist nailed to a plate on a projecting brick corbel. In domestic work, this would have to be concealed by a cornice or coving. Note that the joist is notched over the plate. If a 12 mm hole is drilled first and the notch cut into it as shown, this eliminates the risk of cutting too deep across the fibres and spreads the splitting stress under load over the width of the hole. Fig. 124 shows the use of steel brackets to carry wall plates. These should be built into the wall about every 900 mm to carry a 75 mm by 100 mm wall plate. Fig. 125 shows the joists carried by a wall plate in a brick offset. This can only occur when there is a reduction in wall thickness.

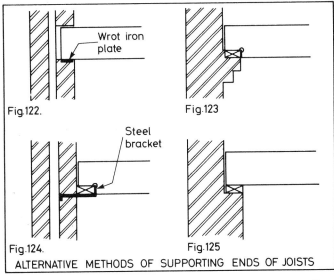

Fig.122.   Fig.123.

Fig.124.   Fig.125.

ALTERNATIVE METHODS OF SUPPORTING ENDS OF JOISTS

The varied forms of stair plans often require a landing which is L-shaped and involves some type of cantilever construction. Fig. 126 shows an example of this. It will be noted that the cantilever joists are carried through the wall of the stairway to another room, the floor of which runs at right angles.

To meet the problem, the wall joist is removed or left out and the cantilever joists from the stairs taken through the walls and nailed down to the second joist. The nailing for the flooring against the wall being made good with noggings cut in.

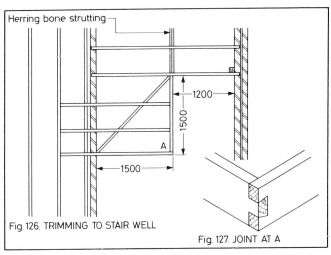

Fig.126. TRIMMING TO STAIR WELL

Fig.127. JOINT AT A

Some herring bone strutting is needed to take the thrust of the landing and this should be countered by diagonal struts cut in as shown. Fig. 127 shows the joint at the external angle.

Finishing the fireplace

Fig. 128 shows the structural detail and finish around a fireplace. To conform to the Building Regulations, as well as meeting the conditions already stated, the hearth should extend at least 150 mm either side of the fireplace opening. The hearth itself should be at least 125 mm thick and of incombustible material; and no combustible material should be under the hearth within 250 mm of the surface, unless there is an intervening air space of 50 mm.

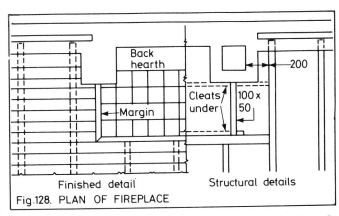

Finished detail   Structural details

Fig.128. PLAN OF FIREPLACE

An exception is made for cleats which support the hearth and ceiling. There are various ways of meeting these requirements and giving the required finish and fixing for the floor boards and mitred margin. One is to use a precast hearth supported by the back of the fireplace opening and a cleat nailed to the trimmer and to bed fixing fillets for the floor boards into the screed for the tiles.

In Fig. 129, asbestos wall board, which may be left in, is

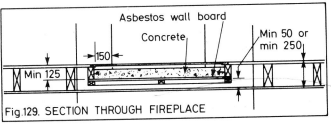

Fig.129. SECTION THROUGH FIREPLACE

carried by cleats nailed to the trimmer and the wall to form the soffit to the concrete hearth poured in situ. The ends being stopped off by 100 mm by 50 mm blockings with nails driven into the sides to give a key into the concrete. These carry the boards and margins at the sides.

Joists on steel beams

Where large floors are to be supported by binders and girders as already described, these structural members are nearly always steel joists. Where the ends rest on the walls or brick piers, stone or concrete pads are necessary to spread the load while binders are fixed to girders by bolted steel angles. The concern of the carpenter however is to fit the joists to the steel beams and also case these in where they project below the floor.

The junctions between the timber joists and the binders may be formed in various ways. The depth of the binder may vary according to the span and loading and may be shallower than the joists it carries. In this case the joists may be notched around the beam as in Fig. 130.

There should be at least 50 mm of timber above the top of the beam to give a nailing for the floor boards. This will not be sufficient to carry any load and the bottom notch of the joist should rest firmly on the bottom-flange of the beam with a gap over the top flange of about 3 mm so that subsequent shrinkage will not transfer the load and cause splitting. A plywood pattern should be made and fitted to the flanges and used to mark out the joints, remembering when cutting, that the whole of the pencil mark is in the waste wood and must be cut away.

When the beam is deeper this construction may not be possible and the beam will project below the ceiling. In this case the exposed part may be cased in one of various ways. Fig. 131 shows the use of plywood casing. The joist is notched around the top flange of the beam and is also supported by timber bearers, or alternatively, steel angle, bolted to the web of the beam. Here again some shrinkage may affect the load distribution at the joist end and a slight clearance should be left over the top flange. Vertical soldiers are cut into the bottom half of the beam at about 600 mm intervals to carry the plywood casings which should have solid angles as shown.

When sound insulation becomes paramount, as in flats and hotels, the floor may be constructed as in Fig. 132 which shows a separate ceiling carried on the bottom flange of the beam. The sound insulation is improved if 50 mm of sand is laid direct on the plaster or board ceiling.

When the span of the building is great enough to require framed floor construction as in Fig. 133, the girder may be much deeper and show considerable projection below the

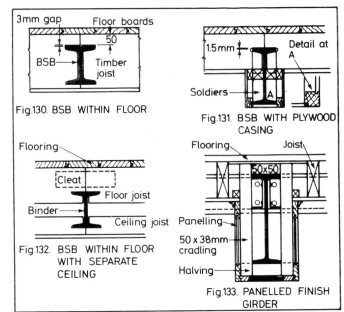

Fig.130. BSB WITHIN FLOOR

Fig.131. BSB WITH PLYWOOD CASING

Fig.132. BSB WITHIN FLOOR WITH SEPARATE CEILING

Fig.133. PANELLED FINISH GIRDER

ceiling. If this occurs in a public building, it may be finished to match panelling on the walls. The drawing shows a section through a girder with a panelled finish, the panelling being fixed to cradling of frames halved together around the girder at intervals of about 600 mm.

Alternatively the panelling can be fixed to vertical soldiers cut to fit accurately between the top and bottom flanges and driven in hard. If a plywood templet is cut which fits hand tight to within 6 mm of the centre web, soldiers marked and cut to this will be a suitable drive fit.

Provision for services

One of the requirements in the construction of suspended timber floors is that of providing grooves or slots in the joists to take the various services without weakening the joists. The maximum bending moment in any beam uniformly loaded (as a joist) occurs at its centre. At the extreme ends, it is subject to shear. Between these positions, notches can be cut in the joists to the depths and within the limits shown in Fig. 134.

Fig. 134 shows limits for cuts in joists to conform to the Building Regulations. Holes should be of a diameter not more than 1/4 the depth of the joist. They should be on the neutral axis and should be not less than 3 diameters apart (centre to centre). Individual slots should be made for each pipe or cable with spaces between to allow for the nailing of floor boards. A hole should be drilled through the joists at the required depth and the notch sawn into it. This is shown on the left of Fig. 135, the wrong method being shown on the right. Note here that the saw has been taken too deep (a very easy mistake to make).

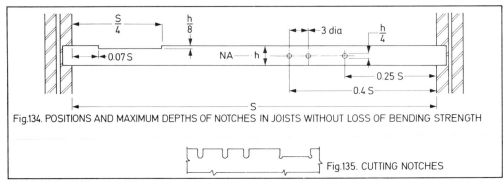

Fig.134. POSITIONS AND MAXIMUM DEPTHS OF NOTCHES IN JOISTS WITHOUT LOSS OF BENDING STRENGTH

Fig.135. CUTTING NOTCHES

Timber, being a natural material, will vary in its strength and in the disposition of its weaknesses from one piece to another. When laying joists in position, the carpenter should watch for large margin knots and other weaknesses and see, if possible, that they are on the tops of the joists. All horizontal beams should be laid with the camber, if any, upwards. Where some timbers appear to be weaker than others, these should be distributed evenly throughout the floor and not all put together in one place.

Floors outside the scope of the Building Regulations

As already stated, when a structural floor does not come within the category of up to three storey domestic buildings privately occupied, the deemed-to-satisfy examples in the Building Regulations do not apply. The design of the floor must then be based on calculations conforming with the requirements of BS 5268 part 2 as to quality of timber, moisture content, safe fibre stress, modulus of elasticity, compression and shear.

For example, given the span of the joists $= 4\,k$ and spacing $= 600\,\text{mm}$, load $= 2\,\text{kN/m}^2$, then from BS table 8 for strength class SC4 $\sigma_{m,adm\parallel} = 7.5\,\text{N/mm}^2$ $\sigma_{c,adm\parallel} = 7.9\,\text{N/mm}^2$, $\sigma_{c,adm,\perp} = 2.4\,\text{N/mm}^2$, $\tau_{adm\parallel} = 0.81\,\text{N/mm}^2$ and $E_{mean}\,9900\,\text{N/mm}^2$. Then accepting each type of stress as being represented by the basic symbol, according to the type of stress being dealt with the joist can be designed for bending and then checked for deflection and shear.

For a simply supported beam with uniformly distributed load the bending moment equation is $\dfrac{FL}{8} = \dfrac{\sigma bh^2}{6}$ which can be adjusted to $bh^2 = \dfrac{6FL}{8\sigma}$.

$F = 4 \times 0.6 \times 2 \times 10^3\,\text{N} = 4.8 \times 10^3\,\text{N}$ $L = 4 \times 10^3$
$\sigma = 7.5\,\text{N/mm}^2$.

By substitution.
$$bh^2 = \frac{6 \times 4.8 \times 10^3 \times 4 \times 10^3}{8 \times 7.5} = 1920000\,\text{mm}^3$$

Let $b = 50\,\text{mm}$. Then $h = \sqrt{\dfrac{1920000}{50}} = 196\,\text{mm}$.

Check for deflection. Deflection formula is $D = \dfrac{5FL^3}{384EI}$

$$I = \frac{50 \times 200^3}{12}.$$

By substitution $D = \dfrac{5 \times 4.8 \times 10^3 \times 4000^3}{384 \times 9900 \times 3.3 \times 107} = 5.1\,\text{mm}.$

Permitted deflection $= 1/300\ \text{span} = \dfrac{4000}{300}\ 13\,\text{mm}$ so this is satisfactory.

The amount of end bearing is generally specified as 75 mm, but calculated safe bearing

$$= \frac{\text{Load on joist end (N)}}{\text{compression perpendicular to grain N/mm}^2}$$

$$= \frac{2.4 \times 10^3}{2.4 \times 50} = 20\,\text{mm}.$$

The generally recommended bearing of 75 mm is therefore usually satisfactory.

Taking the shear strength of the joist, the equation is

$$V = \frac{2bh\tau}{3}.$$

In terms of joist size $bh = \dfrac{3V}{2\tau} = \dfrac{3 \times 2.4 \times 10^3}{2 \times 0.81} = 4444\,\text{mm}.$

Taking b as 50 mm $h = \dfrac{4444}{50} = 89\,\text{mm}$ which is satisfactory.

If the strength of trimmers or trimming joists is in question, this can be dealt with in the same manner as ledgers in the chapter on basic formwork design, reducing the effective breadth of the member to allow for substance cut away to make joints.

CHAPTER 4

Centres and Arches

The use of arches to span openings in this country dates from Roman times. Although steel and reinforced concrete provided a cheaper and generally more economic alternative, arches are still essential to buildings of certain styles and construction, especially those with brick or stone facings.

Before any arch can be erected, it is necessary to provide temporary timber framing, known as a centre, which serves the dual purpose of providing the correct outline and supporting the voussoirs (bricks or stones forming the arch) until the completed arch is ready to carry its load.

SETTING OUT THE ARCH OUTLINE

The first essential for the carpenter is to be able to set out the arch outline, not only to scale on the drawing board, but also full sized on a large floor or platform, under conditions which do not foster drawing board techniques.

There is a variety of curves on which arches may be based, but those used are mainly the circle, the ellipse and the parabola. The setout of some of them is illustrated and described here.

The semi-circle should provide no problem. The centre point from which the curve is struck being at midspan on the springing line.

The segmental arch has its centre below the springing and is easily set out with T-square and compasses by bisecting between two points on the curve (usually at the crown and springing) and producing to meet the centre line as in Fig. 138.

The flatter the arch, however, the longer the radius with an increased proneness to inaccuracy and inconvenience at full scale. With the ready availability of pocket calculators, together with the convenience of the SI system, the precise radius can be worked out in less time than it takes to do the job graphically.

Calculations are based on the property of the circle that if any two chords of the circle intersect, then the product of the numerical values of the two parts of the one equals that of the numerical values of the two parts of the other.

Thus in Fig. 136 $ab \times bc = db \times be$. If these two chords contain the springing and rise of the segmental arch, then the two parts of one will each be half the span and the two parts of the other the rise, and the diameter minus the rise.

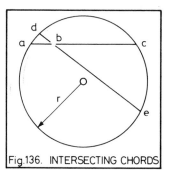

Fig.136. INTERSECTING CHORDS

Thus in Fig. 137,

$$h.X = (\tfrac{1}{2}s)^2 \text{ so } X = \frac{(\tfrac{1}{2}s)^2}{h}$$

$$X + h = \text{the diameter so that radius} = \frac{\dfrac{(\tfrac{1}{2}s)^2}{h} + h}{2}$$

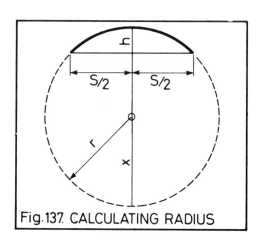

Fig.137. CALCULATING RADIUS

EXAMPLE 1: Given that the span of the arch $= 3000$ mm and rise $= 900$ mm then

$$\text{radius} = \frac{\dfrac{1500^2}{900} + 900}{2} = 1700 \text{ mm}.$$

If the proportions or size of the arch curve do not allow it to be struck from a centre, then the curve may be drawn through predetermined points which may be obtained either by calculations or graphically.

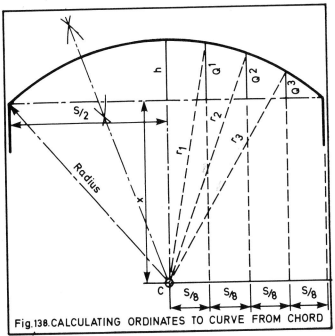

Fig.138. CALCULATING ORDINATES TO CURVE FROM CHORD

The mathematical method uses the theorem of Pythagoras for the right-angled triangle and is illustrated by Fig. 138. From the left-hand side of the drawing it will be seen that $h = r - X$.

$$X = \sqrt{r^2 - \left(\frac{s}{2}\right)^2}.$$

Having obtained the value of X, the lengths of the ordinates Q_1, Q_2 and Q_3 may be found by obtaining the heights of the three dotted right angled triangles and deducting[3] X from each.

Then by the rule of Pythagoras:

$$Q_1 = \sqrt{r^2 - \left(\frac{s}{8}\right)^2} - X$$

$$Q_2 = \sqrt{r^2 - \left(\frac{2s}{8}\right)^2} - X$$

$$Q_3 = \sqrt{r^2 - \left(\frac{3s}{8}\right)^2} - X.$$

EXAMPLE 2: Given span $= 2500$ mm and radius $= 3000$ mm.

Then $X = \sqrt{3000^2 - 1250^2} = 2727.2$ mm.

Then $h = 3000 - 2727.2 = 272.8$ mm.

$$Q_1 = \sqrt{3000^2 - \left(\frac{2500}{8}\right)^2} - 2727.2 = 256.5 \text{ mm.}$$

$$Q_2 = \sqrt{3000^2 - \left(\frac{2 \times 2500}{8}\right)^2} - 2727.2 = 207 \text{ mm.}$$

$$Q_3 = \sqrt{3000^2 - \left(\frac{3 \times 2500}{8}\right)^2} - 2727.2 = 122.6 \text{ mm.}$$

Fig. 139 shows the graphic method of obtaining points in the curve. The procedure is as follows:

1. Set out the span and rise and draw diagonals as shown.
2. Draw perpendiculars to the diagonals to intersect the horizontal tangent at the crown.
3. Divide each half of the tangent, and the springing line and also the vertical height into the same number of equal spacings.
4. Draw intersecting inclined lines in both directions to give the progressive points in the curve. These may be joined up as before.

To draw a normal to a circular curve when the centre is not available, mark off points equidistant from P and from these with compasses draw a bisector.

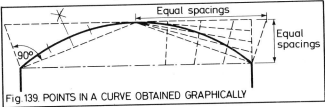

Fig.139. POINTS IN A CURVE OBTAINED GRAPHICALLY

Arches which terminate in a point, as in Figs. 140 to 142, are termed Gothic arches and are sub-titled Equilateral, if the curve centres are at the end of the springing as in Fig. 140, Lancet, if beyond the arch as in Fig. 141 and Depressed, if within the arch as in Fig. 142.

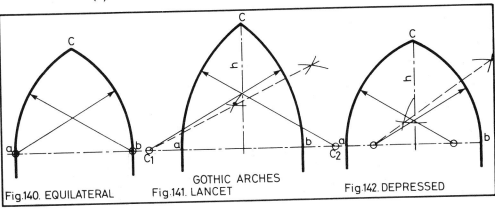

GOTHIC ARCHES

Fig.140. EQUILATERAL   Fig.141. LANCET   Fig.142. DEPRESSED

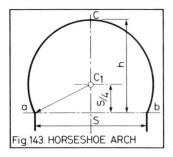

Fig.143. HORSESHOE ARCH

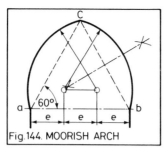

Fig.144. MOORISH ARCH

The construction of Fig. 140 is self-evident, but Figs. 141 and 142 are drawn by setting out the span 'ab' and the rise 'h' and bisecting the points 'ac' or 'bc' to intersect the springing line, extended where necessary.

Other arches with a somewhat foreign influence are shown in Figs. 143 to 146. They are more decorative than scientific. In Fig. 143, the horseshoe arch is drawn from a centre point 'C1' taken at 1/4 of the span above the springing.

In Fig. 144, an equilateral triangle is drawn, the base divided into three equal parts by verticals which are intersected by bisectors of the sides to between points 'b' and 'c' to give the centres for the curves.

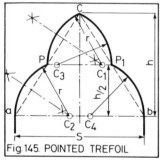

Fig.145. POINTED TREFOIL

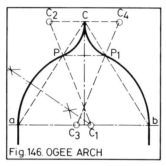

Fig.146. OGEE ARCH

Fig. 145 shows a pointed trefoil. The span and rise are set up and an isosceles triangle 'abc' constructed. A horizontal 'PP1' is then drawn at mid-height, 'CP' is then bisected to give the centre 'C1' for the top curve and 'aP' bisected to give the centre 'C2' for the bottom curve, repeating about the centre line to obtain centres 'C3' and 'C4'.

Fig. 146 shows an ogee arch. Given the span and rise, the relative radii of the convex and concave curves may be adjusted at will. The end of the lower curve in each side is decided on and a horizontal line 'PP1' is drawn; 'aP' is then bisected to give the centre of the lower curve on the springing at 'C1'. The line 'C1P' produced to meet a horizontal at 'C' gives the centre of the top curve at 'C2' and 'C2PC1' becomes a common normal to both curves; 'C3' and 'C4' are marked off horizontally equidistant from the centre line.

Note the reasoning behind this construction. As lines 'C2C4' and 'ab' are parallel, the lines 'aC' and C1C2' intersect at 'P' to give similar isosceles triangles with apexes at 'C1' and 'C2'. They must, therefore, each accept circular curves drawn from these points through their ends.

The second most common curve is probably the ellipse. A true ellipse, however, has several disadvantages from the point of view of setting out and construction in relations to arches. Where parallel widths are involved, such as occur in arch outlines and possibly enclosed ribs to joinery within the opening, only one of the curves can be a true ellipse. Parallels to this will have no geometric significance

and can only be drawn through points measured off the ellipse.

Every normal necessary to joints in the voussoir must be separately set out and every voussoir in the arch from the springing to the crown will be a different shape. For these reasons an approximation, of compound circular curves, is generally substituted for the elliptical form, the name given being related to the number of compass points involved.

Fig. 147 shows a three centre, approximately elliptical arch which is very near to the true shape provided the arch is not very shallow.

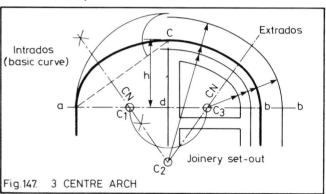

Fig.147. 3 CENTRE ARCH

The springing line 'ab', the rise 'dC' and the diagonal 'aC' are first drawn and a quadrant drawn from centre 'd'. The difference from point 'C' is taken down on to the diagonal and the lower part bisected to cut both the springing and centre line produced in points 'C1' and 'C2' respectively. The bisector is then common normal and is repeated on the other side of the diagram as shown. The curves at the haunches are drawn from C1'' and 'C3' and that at the crown from 'C2'.

It is worth commenting that, where there is a brick arch over and a framed door or window under the intrados of the arch, a number of parallel curves will have to be drawn. Their inter-relation involves co-ordination between the bricklayer, the carpenter who constructs the centre and the joiner who sets out the door or window so that the same centres may be used for all related curves. The most significant outline and the one therefore with which all setting begins is the intrados of the arch.

A five centre arch gives a very close approximation to the true shape, as the two points initially plotted on either side of the crown would actually be points on the true ellipse. Details are shown in Fig. 148. To avoid confusion, the left-hand side is confined to plotting points 1 and 2 in the curve; while the right-hand, starting with these points, illustrates the method of finding the centres and drawing the curves.

Taking the left-hand side first, the major and minor axes as of a true ellipse are drawn first, with the containing rectangle. The semi-major axis and the end vertical are each divided into three equal parts, radial and converging lines are then drawn as shown to intersect and give the required points in the curve. These are then transferred to the right-hand side horizontally and equidistant from the minor axis.

'CP1' is then bisected and produced to cut the extended centre line at 'C1' which is the centre for the crown curve. A line drawn from 'P1' to 'C1' gives the common normal to contain the centre for the next radius. Space 'P1P2' is

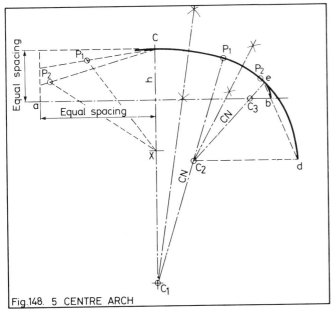

Fig.148. 5 CENTRE ARCH

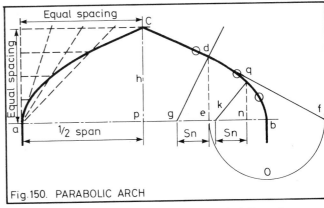

Fig.150. PARABOLIC ARCH

next bisected to cut the common normal at 'C2' giving the next centre. A horizontal is drawn from 'C2' and the arc carried down to meet it at 'd'. A line is then taken back from 'd' through 'b' to cut the curve in 'e'; 'e' is then joined back to 'C2' cutting the springing at 'C3' giving the common normal and centre for the end curve.

Centres for the left-hand side are simply obtained by projecting horizontally to points equidistant from the centre line. The simplest way of drawing a true ellipse is shown in Fig. 149. The major and minor axes are set out and a trammel of a conveniently stiff material marked off with the lengths of half the major and minor axes from the same end. Any number of points in the curve may be marked off from the end of the trammel, each time placing the short point on the major axis and the long point on the minor axis.

The curve is best drawn by pencilling smoothly around a lath bent to coincide with three or more points in the curve. Normals are drawn by bisecting angles formed between lines joined back to the focal points. Focal points are cut off on the major axis at a radius of half the major axis from the end of the minor axis.

The parabola is another curve which is followed more or less closely in forming arch outlines. A true parabolic arch is most likely to be used in masonry construction. It is shown, in outline in Fig. 150. It is not in itself a parabola in the sense that it is halved on the original vertical axis and the parts reassembled at right angles so that the parabola rise becomes half the span while the arch rise with the point at the crown is formed by bringing together the two half bases. The left-hand side shows the method of plotting points in the curve and is self-explanatory while the right-hand shows the method of obtaining normals to the curve.

Taking any convenient point 'd', a perpendicular is dropped from it on to the base at 'e', the distance 'eb' being repeated to the right of 'b' at 'f'. Then 'f' joined to 'd' becomes a tangent at the point, so that a line 'dg' drawn at right-angles becomes a normal. The distance 'Sn' (subnormal) becomes a constant value in constructing any other normals to the curve. Thus, to draw a normal from a point 'q', draw the vertical 'qn', and measure off the distance 'Sn' to 'k' when 'kg' becomes another normal.

A close copy of the parabolic arch is the Tudor or four centre arch. Fig. 151 shows the method of setting this out.

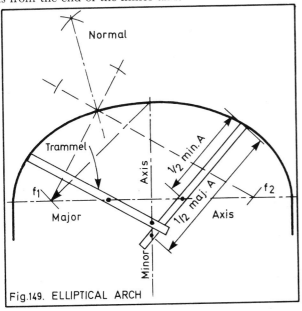

Fig.149. ELLIPTICAL ARCH

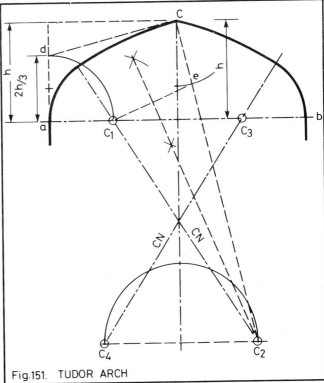

Fig.151. TUDOR ARCH

CENTRES AND ARCHES

Referring to the left-hand side, mark out the span and rise and draw the containing rectangle. Mark off 2/3 of the height at 'd' and join to 'C'. Draw 'Ce' at right angles to 'dC' and produce indefinitely. Mark off 2/3 from 'a' to 'C1' and from 'C' to 'e' and join 'C1' to 'e'. Bisect 'C1–e' and produce to cut 'C–e' extended at 'C2'. Join 'C2' through 'C1' and extend to form the common normal. The centres of the lower and upper curves are 'C1' and 'C2' respectively. The centres 'C3' and 'C4' are horizontally opposite 'C1' and 'C2' and the same distance from the centre line.

CONSTRUCTING AN ARCH CENTRE

In constructing an arch centre the main requirements are:
  1. To provide the correct shape for the arch soffit or intrados.
  2. To give sufficient width of support within the thickness of the wall to give stability to the voussoirs.
  3. To make the centre strong enough, but with maximum economy to carry the superimposed load of brickwork without deflection or failure until the arch is capable of taking over.
  4. To be able to remove the centre gently without lifting or vibrating the green brickwork.

The simplest type of arch is the flat or soldier arch, the soffit of which must appear to be straight, but must, in fact, have a camber of about 1/100 span to correct the illusion of sagging. The centre for thin walls is merely a flat timber beam of suitable width to give support and with the top surface slightly cambered as required. It is referred to as a turning piece and is carried at the ends by props, as described later.

If the arch curve is a flat segment as in Fig. 138 and the rise is limited to within commercially obtainable widths of softwood timber, then a solid turning piece can still be used for arches in half-brick walls, the top being bandsawn and planed to shape. It will, however, require a heavy beam 75 mm thick with little recovery value. Fig. 152 shows an alternative construction.

A rectangular timber beam of sufficient strength to carry

the arch has thin boards nailed to each side cut to the required arch curve, packings being nailed at each end to give stability. In the example given the beam is 100 by 50 mm made up to 80 mm thick by the face pieces. The face pieces could be cut from off-cut plywood.

The centre will need supporting by props at each end. These can be cut to neat length and carefully eased off the wall to strike for small arches. There is, however, some tendency to lift due to the bottom-inside corner binding (see Fig. 154 top). If, however, the bottom of the strut is kept its own thickness off the wall, as in Fig. 154 bottom, the joint being slightly splayed to suit; then no matter how thick or short it may be, it will always clear on striking. It will be easy to force out with a nail bar, but must be strutted against buckling at mid-height. See Figs. 152 and 153.

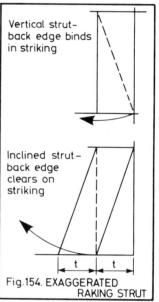

Fig.154. EXAGGERATED RAKING STRUT

Figs. 155 and 156 are elevation and section of a semi-circular centre to an arch in a half-brick wall. The weight is carried by central struts of 150 by 25 mm, the thickness being made up to 75 mm by face pieces and ties nailed to both sides. The usual method of supporting small centres is shown, with posts and folding wedges for adjustment

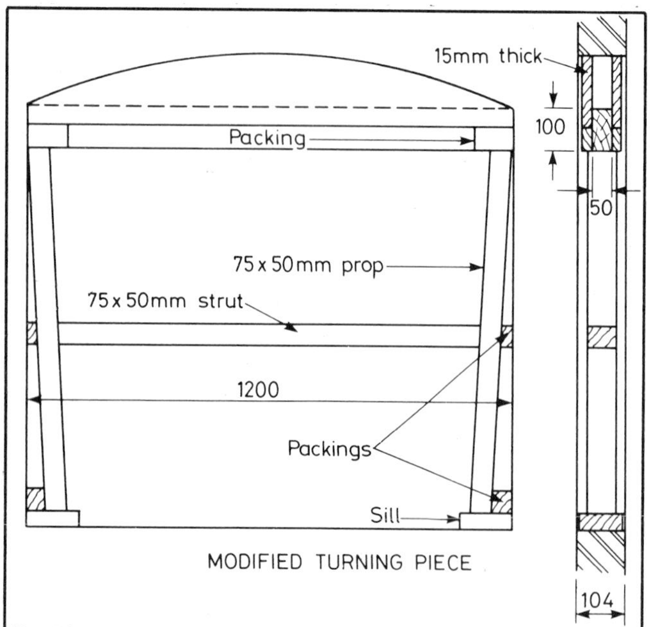

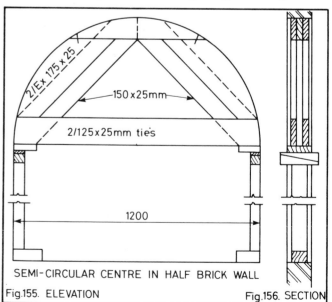

MODIFIED TURNING PIECE

Fig. 152. ELEVATION   Fig.153. SECTION

SEMI-CIRCULAR CENTRE IN HALF BRICK WALL

Fig.155. ELEVATION   Fig.156. SECTION

easing and striking. The posts should be strutted back as in the previous example.

It is also important that centres in thin walls should be given ample lateral support externally by bracing back to some convenient anchorage such as existing walls or floors.

Arches in thicker walls

When walls are over a half-brick thick, twin frames or turning pieces become necessary; these are connected by cross members which provide support for the voussoirs and are known as laggings. Bricklayers and stonemasons have differing requirements in this respect.

Figs. 157 and 158 show elevation and section of a centre to a semi-circular arch in a 1½-brick wall. The design of each truss and the number of curved ribs required depends upon or may govern the width of timber used. They should not be less than 100 mm wide at their ends to give suitable abutment and this, according to the radius of curvature, will govern their lengths.

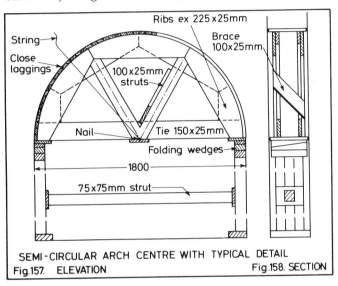

SEMI-CIRCULAR ARCH CENTRE WITH TYPICAL DETAIL
Fig.157. ELEVATION   Fig.158. SECTION

Each truss is made up of two layers of ribs jointed mid-way to each other with struts nailed to the horizontal ties and the exposed surfaces of the inner ribs. All butt joints should be well-fitting to avoid deflection under load. A diagonal brace as shown in section and elevation increases the lateral stability of the centre.

The end supports consist of double sets of posts tied together by heads and sills with intermediate diagonal bracing (not shown). Adjustment is again by means of folding wedges. These fit between the post heads and cross-bearers nailed to the centres and may be greased for ease of striking.

The overall width of the centre over the laggings should be 12 mm less than the thickness of the wall so as not to foul a line or straight edges. The trusses should be set 25 mm in from the ends of the laggings to reduce the risk of splitting them when nailing.

In Fig. 157 the laggings are close fitting and should be planed off to give a smooth, accurate surface suitable for a gauged brick arch with fine joints. An alternative is to use one or more layers of plywood bent and nailed around the ribs as in Fig. 159. Where not such a fine finish is required as for a rough or axed brick arch, open laggings about 18 mm apart (Fig. 159 top) are satisfactory.

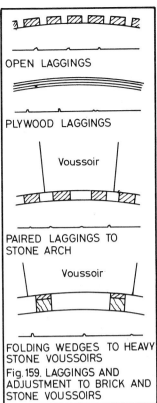

Fig.159. LAGGINGS AND ADJUSTMENT TO BRICK AND STONE VOUSSOIRS

In masonry arches, the voussoirs are generally much bigger and it is customary to set out the laggings two to each voussoir (Fig. 159 third down). When voussoirs are very large, the mason likes to be able to set and adjust each one. He prefers to do this directly off the tops of the ribs with folding wedges (Fig. 159 bottom). Where possible the bricklayer likes to have a nail in the centre or below it at the centre of each curve so that he can make each brick truly radial. The stonemason likes to have a radius rod similarly centred so that he can check his soffit.

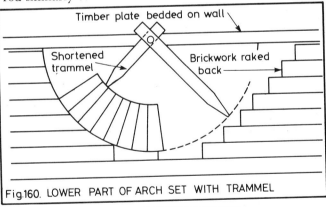

Fig.160. LOWER PART OF ARCH SET WITH TRAMMEL

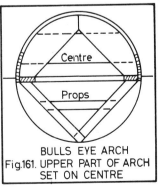

BULLS EYE ARCH
Fig.161. UPPER PART OF ARCH SET ON CENTRE

41

CENTRES AND ARCHES

A special type of arch is the bull's eye shown in Figs. 160 and 161. This involves two separate operations. The bottom half of the ring of brickwork is built up and raked back as in Fig. 160 right. It is then made up to the required circular curve against a trammel centred on a bedded horizontal plate. The trammel is then shortened and the lower voussoirs set as in Fig. 160 left. The semi-circular centre constructed with laggings in the usual way is then fixed and the top half of the arch set.

Rib frames for other arches

Figs. 162 to 164 show the framed assembly for three other shapes in arch centres, the rest of the construction being as in Figs. 155 or 157.

Fig. 162 shows a small Gothic centre. As only two ribs are needed, they form a stable triangle in effect and only need tying at the top and bottom.

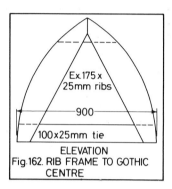

Fig.162. RIB FRAME TO GOTHIC CENTRE

Fig. 163 shows the rib frame for a three centre arch. It should be noted that there is no necessity to make the joints on the common normal. In setting out the rib joints, too much cross grain should be avoided and the end of any rib should not be less than 100 mm wide. The best way to set out or arrange the ribs in any centre is first to draw two parallel curves inside the rib outline, the first being 100 mm less and the second the available plank or board width less. This is shown on the left-hand side of Fig. 163.

Then the inner line of each rib should be inside or tangent to the inner curve, but should not cut through the outer one. Note also that when the ribs are built up in two layers, it is an advantage to have sufficient of the inner rib exposed to provide a nailing for the strut.

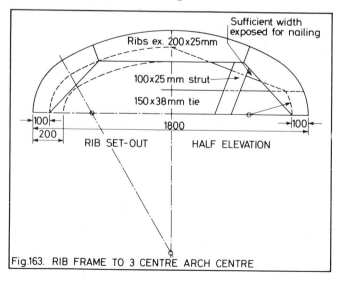

Fig.163. RIB FRAME TO 3 CENTRE ARCH CENTRE

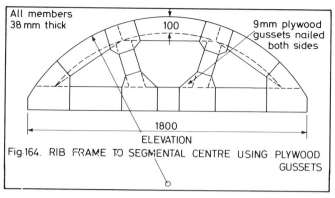

Fig.164. RIB FRAME TO SEGMENTAL CENTRE USING PLYWOOD GUSSETS

A segmental arch centre with an alternative construction is shown in Fig. 164. Here all the timbers are of the same thickness, are butt-jointed together and are single only. All the joints are connected by plywood gussets nailed both sides.

The load of brickwork on an arch in a brick wall is said to be only taken from a triangle within the overhang of the bonded courses on either side. I am doubtful of this, at least while the mortar is in a plastic state; but in most cases the load is likely to be limited. If, therefore, the centre is made with well-fitting joints and robust enough to be roughly handled, it is not likely to fail in use.

Where complicated geometrical construction is involved in setting out arches, the centre may first of all be set out to a scale of, say, 1 : 5; from this drawing; the positions of centres can be plotted full size.

LARGER ARCHES

In the construction of large arch centres, the loading becomes correspondingly heavier; consequently more consideration must be given to ensure that the erection is sound in design principle and construction. The actual loading on an arch centre depends upon a number of factors, such as the shape of the arch (the flatter the arch, the heavier the load), the rate of building, and the wall thickness. The design of the centre itself must also depend upon whether or not intermediate support is available or the full load has to be taken at the abutments.

Generally speaking, the design of arch centres is largely a matter of common sense with the application of certain broad principles including the following:

1. Triangulation: where possible the centre lines of structural members should be arranged into triangles with loadings concentrated at the node points; so putting all members into direct tension or compression. It also stiffens the truss so that it will not be liable to distort during handling, or when subject to unequal pressures during intermediate stages of loading.
2. Adequate fixing at the joints with accurate fitting and close contact on all bearing surfaces.
3. Transverse stability achieved by adequate bracing and lateral ties between the trusses.
4. Sufficient stiffness in long struts to avoid the risk of buckling under load.
5. The design of joints in tension to avoid failure by shear.

The actual width of the timber available for the arch

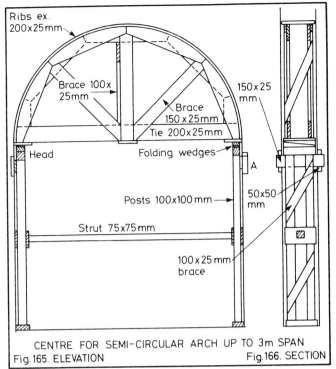

CENTRE FOR SEMI-CIRCULAR ARCH UP TO 3m SPAN
Fig. 165. ELEVATION   Fig. 166. SECTION

as shown in section in Fig. 166. Transverse bracing, plus the 75 by 75 mm strut at mid-height, prevents the supporting posts from buckling either way.

Fig. 167 shows a centre for a semi-circular arch about 4.5 m span. The length 'A' is obtained by drawing curves of radii 100 mm and 225 mm respectively less than the rib outline and constructing a tangent to the latter, cut off by the intersections with the former. It will be seen that a minimum of five ribs will be required in this case. The increased span coupled with the greater number of joints will tend to make the centre flexible and unstable and liable to buckle under load.

To increase the rigidity, a simple horizontal truss at section A-A as shown in Fig. 169 has been placed at mid-height. The ties are also braced as in section B-B Fig. 170. These, interconnected with the transverse bracing to the raking members in Fig. 168, add up to an overall firmness sufficient to counter any tendency to buckle under load or fail during erection.

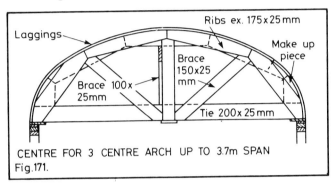

CENTRE FOR 3 CENTRE ARCH UP TO 3.7m SPAN
Fig. 171.

ribs governs the length according to the radius of curvature so as to allow, say, 100 mm width at each narrow end. The increase in width of the rib due to the centre swelling to the arch curve generally gives sufficient strength to take care of any secondary moments.

Fig. 165 shows a centre for a semi-circular arch of about 3 m span. Using 225 by 25 mm timber for the ribs, four will be needed for the outer ribs, the joints on the inner ones coming halfway between them.

Three braces will be needed, although they will act chiefly as struts. Assuming the wall is two bricks thick some transverse rigidity can be given by internal braces

Fig. 171 shows the elevation of a three centre, approximately elliptical arch centre. The joints have not been placed on the common normal, but have been arranged so that all the ribs are about the same length. This enables the struts to meet at mid-span on the tie without the raking

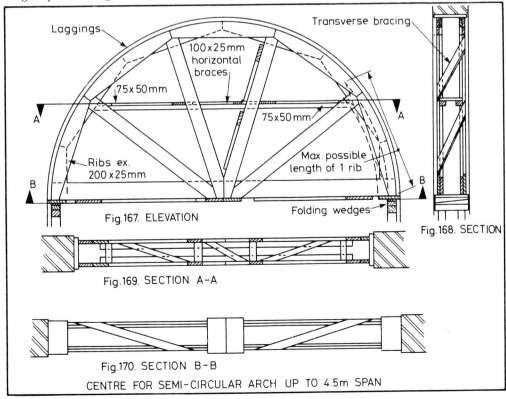

Fig. 167. ELEVATION

Fig. 168. SECTION

Fig. 169. SECTION A-A

Fig. 170. SECTION B-B

CENTRE FOR SEMI-CIRCULAR ARCH UP TO 4.5m SPAN

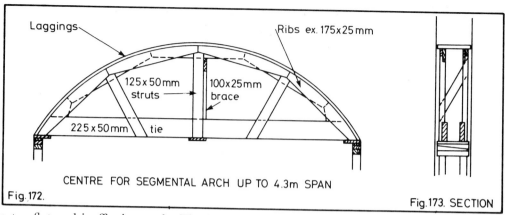

Laggings

Ribs ex. 175x25 mm

125 x 50 mm struts

100x25mm brace

225 x 50 mm   tie

CENTRE FOR SEGMENTAL ARCH UP TO 4.3m SPAN

Fig. 172.

Fig. 173. SECTION

ones being at a too flat and ineffective angle. The bottom ribs will have to be wider and make-up pieces have been used for this.

Fig. 172 shows a flat segmental arch of up to 4.3 m span, using 175 by 25 mm timber. Only four outer ribs will be needed to each face. The braces or struts are normal to the curve as they would be at too flat an angle if taken to meet at mid-span. This means that the tie will be subject to secondary or bending stresses with point loads from the struts. As the arch is flat, these loads may be considerable. The tie has to be presumed to act as a beam and is made 50 mm thick. The struts are notched halfway on to the tie and also over the inner ribs. Thus the load on each strut is taken by direct bearing from both inner and outer ribs, giving two chances of security against possible poor fitting. The loads are similarly transmitted to the ties by direct bearing.

VERY LARGE ARCHES

Arches with spans above those so far dealt with are more likely to be used in civil engineering structures such as bridges, the breadth measured in metres rather than millimetres. Much heavier loads are therefore likely to be imposed. For greater economy, trusses are spaced about 1.2 m centres. The laggings will virtually act as beams and may have to be 75 to 100 mm thick.

The support system to the centring may be tubular scaffolding or a heavy duty proprietary system such as 'Acrow Shoreload', but as this is 'site carpentry', I have decided to stick to timber supports.

Where possible and convenient, direct intermediate support to the ground should be taken from one or more intermediate points along the span. In this way, all the struts or braces may be in direct compression with loads collected at the heads of the posts and transmitted direct to the ground. The ties will then merely serve to couple members together and provide a fixing for lateral bracing. Smaller timbers may be used, and the joints made simpler.

Figs. 174 and 175 show centring, to a segmental arch to a bridge of about 6 m span, one intermediate support being permitted. Trusses are spaced at 1.2 m centres with 100 mm laggings. All joints should be cut carefully to give full bearing area at end grain and nails staggered to reduce the risk of splitting along the grain. Ribs are subject to bending and compression, struts to compression only. Ties may be subject to some tensional stresses during initial loading at the haunches.

The trusses are adequately tied together by the laggings around the intrados and are supported at the abutments and at the centre by systems of heads, posts, braces and sills. Where the ground is poor, loads from the sills will need to be spread by using, say, hardcore, weak concrete or a sleeper grillage.

The heads are paired with runners tying the trusses together. These, combined with the struts and transverse braces (Fig. 175) to form a rigid overall construction above the springing.

Provision for overall adjustment is made with triple wedges between the paired horizontals. If the outer wedges are tacked to the head and runner respectively, the centre ones, made conveniently longer, can be tapped one way or the other to raise or lower the centring. The joints

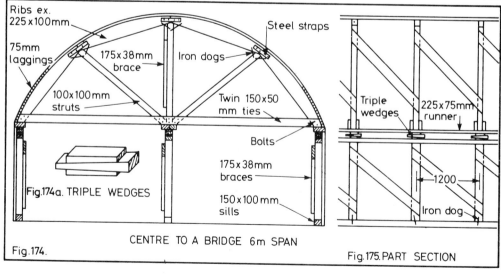

Ribs ex. 225 x 100 mm

75 mm laggings

175 x 38 mm brace

Iron dogs

Steel straps

100 x 100 mm struts

Twin 150 x 50 mm ties

Bolts

Fig. 174a. TRIPLE WEDGES

175 x 38 mm braces

150 x 100 mm sills

Triple wedges

225 x 75 mm runner

1200

Iron dog

CENTRE TO A BRIDGE 6m SPAN

Fig. 174.

Fig. 175. PART SECTION

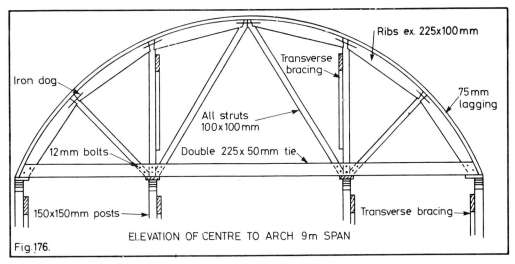

ELEVATION OF CENTRE TO ARCH 9m SPAN

Fig. 176.

between the struts and ribs are secured by means of steel straps bolting the ribs together and iron dogs which have the effect, when driven, of pulling the joints tight to ensure minimum settlement when under load. These fittings also ensure that the trusses will be rigid enough to allow them to be hoisted into position.

Fig. 176 shows a centre to an arch of 9 m span. Details of support and construction are the same as before, but two intermediate rows of supports are provided. For the sake of uniformity of timber sizes, ribs are all made the same length, intermediate posts being directly under alternate joints. Struts are taken through and the ribs shouldered into them. These might have to be built up in situ (having been fitted together on the ground first), temporary light bracing being used to aid assembly.

The system shown gives more positive connections and will make assembly easier under these conditions. As the trusses will not have to be manhandled, the steel straps previously shown are not necessary.

The last drawing, Fig. 177 is the elevation of a centre to an approximately elliptical arch which, perhaps because it is over a river or a railway cannot have any intermediate obstructions. The actual supports, although confined to the abutments and much heavier, are otherwise in principle the same as before, as will also be the transverse bracing between the trusses.

The problems are:

1. To make provision for the heavy load to be carried at the crown.

2. To reduce tensional stresses as much as possible.
3. To have the minimum number of joints, so as to give the maximum degree of stiffness.
4. To confine the timber sizes to those commercially obtainable.

It will be seen in Fig. 177 that the heavy loading at the crown is taken by stout raking struts to the abutments. The struts are notched into the king post which will eventually be under tension. The upper ends of the top ribs take an adequate bearing on the top struts. The still considerable load at the haunches (half way between crown and springing) is also taken by struts which must necessarily be cut against strut 'A'. The joint must have its lateral stiffness increased by 50 mm fish plates bolted on either side.

Between these members are cut intermediates 'B' which also act as struts and carry the curved ribs cut to the arch outline. The upper ribs have make-up pieces; while the lower ones, which will not be very heavily loaded, are packed off the lower struts. The joints between the main struts, the centre king post and the twin tie should be made with split ring connectors as shown.

The ribs will only need to be 75 mm thick, thus reducing cost of material and labour in cutting. Cross bracing between the king posts will be necessary to give stability and resistance to wind. If the ties have to be jointed, this should be done with split ring connectors with 100 mm packings at the joints. Reasonably accurate fitting of all joints is necessary to minimise the risk of settlement under load.

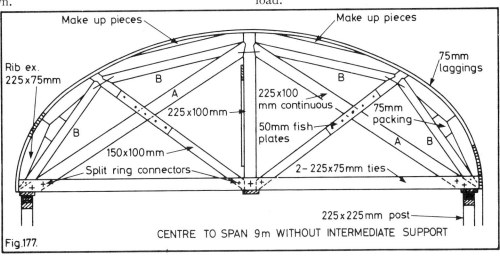

CENTRE TO SPAN 9m WITHOUT INTERMEDIATE SUPPORT

Fig. 177.

CHAPTER 5

Shoring

There are many factors in construction which contribute to the loss of stability in an existing building, with development occurring early or late in its existence.

Among these may be the failure to design the foundations to spread the load suitably over the sub-soil, so that there is an uneven settlement; the effect of frost or drying out of the ground due to shallow foundations; or general weakness in the construction with party walls not suitably bonded to externals, floors not tied to walls or badly built roofs spreading to exert a horizontal thrust on the walls at eaves level.

Local activities also may have an unsettling effect. For example, deep excavations near the wall foundations; the demolition of one house in a terrace; or the removal of part of a wall to take, say, a shop window.

RAKING SHORES

Vertical movement or settlement may be dealt with in anticipation or after discovery by dead shoring or underpinning; but horizontal movement, leading to overturning and ultimate collapse of the wall, must be countered either permanently or until alternative support can be built in by systems of raking or flying shores (according to existing external conditions).

Where support must start from ground level, then pressure must be provided by stout inclined props, known as raking shores, in systems numerically described according to the number of shores or struts required.

The first sign of movement in a building is given by cracks appearing in the walls and is particularly evident in internal plasterwork. These should be watched carefully and if they continue to widen (a piece of stout paper glued over the crack will break to show movement), their cause should be investigated. Where danger is imminent, the wall should be shored up to prevent further movement. The number of shores in the system is generally governed by the height of the wall as evidenced by the number of storeys.

Fig. 178 shows a single raking shore, the basic principles of construction of which apply to more complex systems

also. The shore, which will be subject to compression under load, should be square in section to give equal resistance to buckling both ways unless it may be more closely braced across the narrowest dimension of a rectangular section. The inclined thrust of the raker must be restrained by a vertical reaction in order to exert the necessary horizontal pressure against the wall.

This is achieved by recessing a 100 mm by 100 mm hardwood timber, a 'needle' about 150 mm into the wall and notching the raker around it, the weight of the brickwork thus providing the necessary restraint. The buried end of the needle is reduced to a 75 mm depth (brick course).

So that the inward thrust may be spread over a reasonable area of brickwork, the raker rests against, and the needle is mortised through, a vertical wall piece, about 225 mm by 50 mm to 75 mm section, firmly secured to the wall by wall hooks. A cleat behind the needle stiffens it against the raker thrust. The reaction from the lower end of the shore is spread over a suitable area of ground by setting it on a stout sole piece.

To enable the shore to be tightened without vibration which would be fatal to the presumed already weakened wall, the sole piece is set at an angle of about 80 deg. to the raker. The bottom of the raker is notched to take the end of a crow bar by means of which it may be gently levered in to tighten. Iron dogs driven into raker and sole piece hold the former firmly in position.

Fig. 179 comprises a series of sketches showing constructional detail. It will be noticed that the cleat is taper-housed into the wall piece to offer a positive resistance to the raker thrust. It is necessary to appreciate that no masonry or brickwork which has started to overturn can be pushed back into its original position without creating fresh cracks and further increasing its instability; and also that no appreciable thrust should be exerted against a wall where there is no opposite resistance. To these ends, pressure should be firm and not excessive and should be confined to positions against cross walls or floors.

When a raker is opposed by a floor, three forces are involved. These are the weight of the wall, acting on the needle; the horizontal resistance of the floor; and the line of thrust of the raker. In theory, to avoid shear or bending

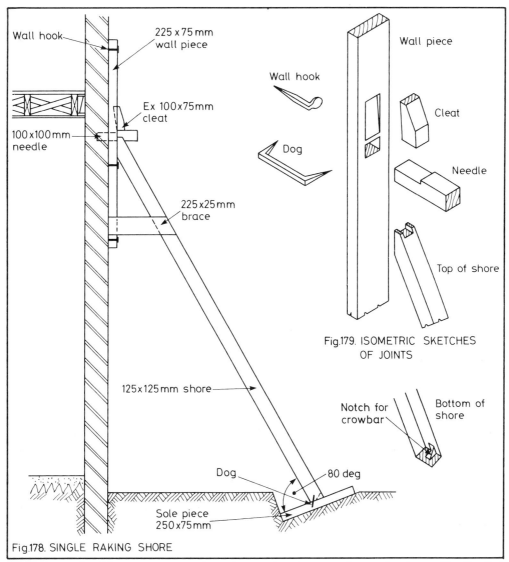

Wall hook

225 x 75 mm wall piece

Ex 100x75mm cleat

100 x 100 mm needle

225 x 25 mm brace

125 x 125 mm shore

Dog

80 deg

Sole piece 250 x 75 mm

Fig.178. SINGLE RAKING SHORE

Wall piece

Wall hook

Dog

Cleat

Needle

Top of shore

Notch for crowbar

Bottom of shore

Fig.179. ISOMETRIC SKETCHES OF JOINTS

stresses in the wall these should meet at a node point, although in practice the stiffness of the wall piece, particularly if it is say 75 mm thick, will take care of inaccuracies of 200 mm to 300 mm.

Fig. 180 shows the force lines where the floor joists are parallel to the wall and pressure resistance from the floor is confined to herring-bone strutting, tightened by folding wedges normally used to stiffen floors. The shoring system should therefore be preferably placed where this stiffening occurs.

Fig. 181 shows a section through a wall with the floor carried on an offset where the thrust of the floor is taken to come through the wall plate to which the joists are nailed. The usual textbook takes the centre of the wall plate as the node point for the thrust of the raker, but in my opinion, unless the floor is loaded near its edge, the bulk of the downward pressure will still come from the wall and the force lines should be taken as shown here.

Fig. 182 illustrates a raking shore against a cavity wall where downward reaction will be on the outside leaf only, while the horizontal resistance will have to come through the wall ties. It would seem therefore that the theoretical lines of thrust having been worked out, the actual centre line of the raker should oppose the nearest wall tie. If the floor joists are bedded direct on the brickwork or rest on a metal bar, and perhaps the plaster joint ceiling to wall has

already opened out, the floor will offer little opposition. In this case, it may be advisable to coach screw a length of scantling to the floor joists as shown.

It is difficult, if not impossible, to estimate accurately the pressures exerted on raking shores when they are levered into position. It will be realised that, as the wall is still standing by itself and must not be pushed back, the only force is that which is applied to steady it, but this must be sufficient to counter any further stresses (unknown). In my opinion, rigidity is more important than great strength in a raking shore system.

A commonsense and practical approach to the problem

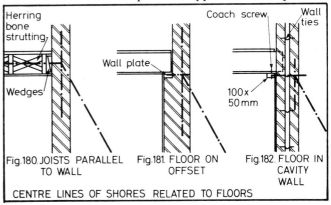

Herring bone strutting

Wedges

Coach screw

Wall ties

Wall plate

100 x 50 mm

Fig.180. JOISTS PARALLEL TO WALL

Fig.181. FLOOR ON OFFSET

Fig.182. FLOOR IN CAVITY WALL

CENTRE LINES OF SHORES RELATED TO FLOORS

will however be helped if the operator has some idea of how the stresses are distributed.

Stress related to raker pitch

Figs. 183 to 187 are force and stress diagrams showing the values involved when rakers are applied at various angles to meet a horizontal standard assumed resistance of 1 kN. Thus, as in Fig. 183, an equal and opposite force of 1 kN would be required.

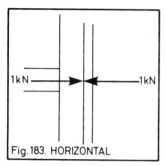

Fig. 183. HORIZONTAL

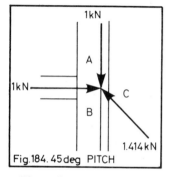

Fig. 184. 45 deg PITCH

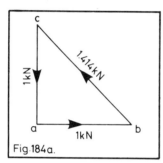

Fig. 184a.

If, as in Fig. 184, the raker were at 45 deg. to the horizontal, the raker would have to exert a pressure of 1.414 kN and would create an upward thrust of 1 kN on the brickwork as shown by the triangle of forces in Fig. 184a.

Figs. 185 and 185a show that, at 60 deg. pitch, a raker will need to exert a pressure of 2 kN and the lift on the brickwork would be increased to 1.732 kN.

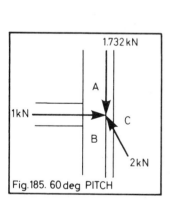

Fig. 185. 60 deg PITCH

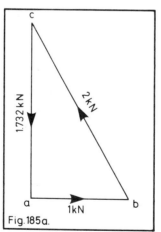

Fig. 185a.

Increasing the raker pitch to 70 deg. the values would be increased to 2.924 kN and 2.747 kN (Figs. 186 and 186a) while at 75 deg. (Figs. 187 and 187a) they would become 3.864 and 3.732 kN respectively. From this it is obvious that a pitch of more than 70 deg. should be avoided if possible.

Where the building is more than one-storey high, a

system of raking shores with one to every floor level is usually necessary. Whether or not support is extended to the roof at eaves level will depend upon circumstances. If the roof is in good condition and well tied back, it may not be necessary.

Fig. 188 shows a multiple raking shore system to a four-storey building. The positions of the heads of the shores are obtained as before. The feet of the shores bed on a common sole piece. They are each levered into position as before with a crow bar, but the angles must become progressively steeper. For the leverage to be effective, the sole piece must be less than 90 deg. to the third raker which will make it about 60 deg. to the inner raker which is fixed and tightened first. The thrust of this will tend to push the sole piece back and a stake may be driven as shown to prevent this. It is better to arrange for the feet of the rakers to be 100 mm to 200 mm apart as this will allow a greater degree of adjustment for tightness. The feet of the shores should each be dogged both sides to the sole piece.

Fig. 189 shows the force lines of a triple shore with values taken from the previous examples converging on a sole piece, the reaction of 6.100 kN and the resistance to sliding of 1.450 kN being taken from the polygon of forces

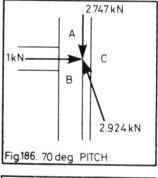

Fig. 186. 70 deg PITCH

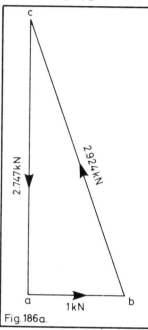

Fig. 186a.

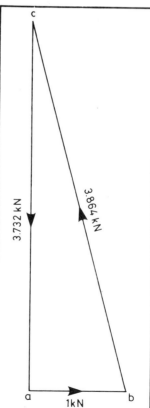

Fig. 187a.

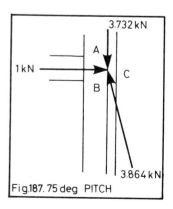

Fig. 187. 75 deg PITCH

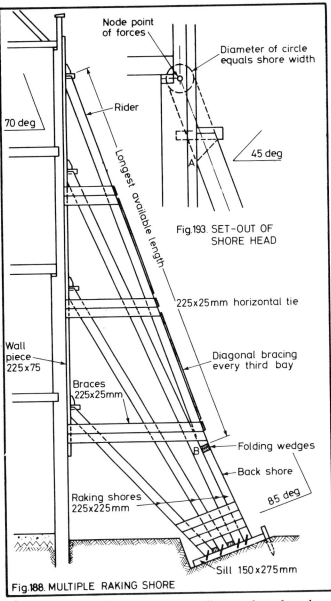

Fig.193. SET-OUT OF SHORE HEAD

Node point of forces

Diameter of circle equals shore width

45 deg

Rider

70 deg

Longest available length

225×25mm horizontal tie

Diagonal bracing every third bay

Wall piece 225×75

Braces 225×25mm

Folding wedges

Back shore

85 deg

Raking shores 225×225mm

Sill 150×275mm

Fig.188. MULTIPLE RAKING SHORE

a 'rider', the height being made up with a shorter member resting on the third raker and called a 'back shore'; the adjustment to this being made with folding wedges. It will be seen that one continuous wall piece takes all the needles.

As the rakers are of great length, it is necessary to brace them with 25 mm boarding. This should be nailed close to the heads of the respective shores to have the maximum retentive effect.

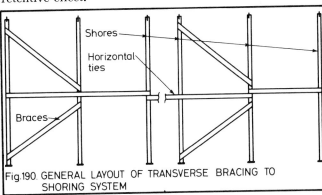

Shores

Horizontal ties

Braces

Fig.190. GENERAL LAYOUT OF TRANSVERSE BRACING TO SHORING SYSTEM

Lateral bracing is shown in Fig. 190 with continuous central ties and a set of braces every third bay. The spacing of the shoring systems will depend upon the wall conditions and the positions of windows, etc., but generally varies between 2.5 to 5 m. If the rakers have to be jointed to length, they are best bolted together with fishplates, as shown in Fig. 191, with well-fitting butt joints. The joints will always present weak places against buckling and should be considered when bracing in either direction. Wall pieces may be halved and nailed as in Fig. 192.

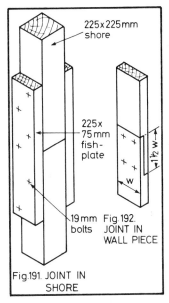

225×225mm shore

225×75mm fish-plate

19mm bolts

Fig.191. JOINT IN SHORE

Fig.192. JOINT IN WALL PIECE

Setting out raking shores and erecting

Having obtained on site the details of the building, floor levels, etc., and the ground width available on plan, the elevation should be drawn to as large a scale as conveniently possible. From this it should be possible to obtain the lengths of the timbers needed. Using a long measuring rod, the holes for the needles may then be marked on the walls to the nearest brick course level and the holes chopped out. The measuring rod is then re-marked to the actual holes and from it the wall piece marked, mortised and housed.

in Fig. 189a. As before they are not direct values, but show the stresses which would be exerted on the sole piece necessary to create pressures of 1 kN at the three floor levels.

In Fig. 188, it is assumed that the excessive length needed for the outer raker prevented its use in one piece. A convenient length is, therefore, taken for the top part, termed

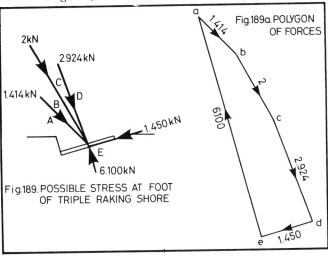

2kN

2.924 kN

1.414 kN

C

D

B

A

1.450 kN

E

6.100 kN

Fig.189. POSSIBLE STRESS AT FOOT OF TRIPLE RAKING SHORE

a

1.414

b

2

6.100

c

Fig.189a. POLYGON OF FORCES

2.924

d

e 1.450

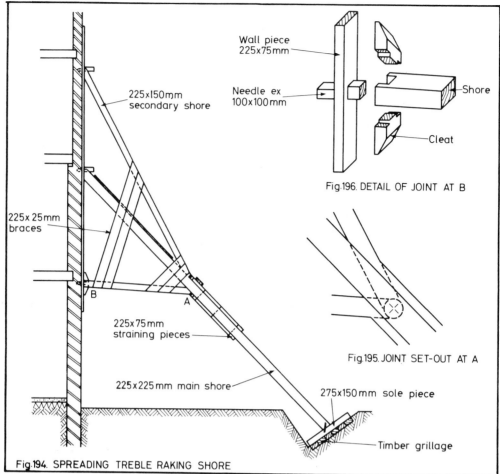

Fig.196. DETAIL OF JOINT AT B

Fig.195. JOINT SET-OUT AT A

Fig.194. SPREADING TREBLE RAKING SHORE

The wall piece is then fixed with wall hooks, the needles driven in and the supporting cleats fixed. A line may then be stretched from the needle for the longest raker position to the measured distance from the wall; the earth is dug out and the sole piece set and firmed to the required angle to the line.

The shortest raker may then be marked out and the head cut, its length checked on site and its foot cut and notched for the crow bar. It may then be erected, levered tight and secured with iron dogs. Other shores may then follow in turn. The rest of the work is straightforward.

Fig. 193 shows a method of setting out the head of the raker to give an equal width of bearing both against the wall piece and the needle and is self-explanatory.

Fig. 194 shows a spreading raking shore to a 3-storey building where there is ample ground space, and allows some economy in timber. The shore to the second floor is set first at 45 deg. to the horizontal.

It is assumed that the ground support is poor and a grillage of sleepers is used to take the thrust from the sole piece. The lines of force of the three raking members should meet at the centre of the middle raker and Fig. 195 shows how this is achieved. Fig. 196 is an exploded detail of the joint at B.

A generally accepted rule is that rakers should be square in section the width of which should be 1/60th of the wall height. Thus if the wall is 12 m high then the raker section should be

$$\frac{12,000}{60} = 200 \text{ mm}$$

so a 200 mm by 200 mm raker will be required. These large sizes are dictated more by the length of the strut than by the load carried. By introducing more intermediate bracing to prevent buckling, sizes could be reduced. However, the extra bracing and the labour in cutting it and fixing it would cost money and a happy medium has to be found.

Large timbers, if not cut about, will have a high reclaim value. There are other practical considerations and 125 mm square is considered about the minimum size for nailing, fitting to needles, etc. Larger timbers fit together more solidly and compress less under load.

Steeply pitched rakers, as well as being less effective, are also a danger in that, if they become a little slack through settlement, they will allow considerable horizontal movement before again taking up their load.

FLYING SHORES

When a building is close to another, then it may be much more effectively supported from that building by systems incorporating horizontal struts together with raking struts at ideal angles of 45 deg. These are known as flying shores numerically described according to the number of horizontal shores contained in the system as single, double, flying shores and so on. The two commonest uses are to support a weak wall to a building from a conveniently close strong one, and to provide mutual support between two party walls in a terrace exposed and weakened by the demolition of one of the houses. Buildings of one or two storeys will only need one or two horizontal struts suitably braced, but above this height a more detailed arrangement becomes necessary.

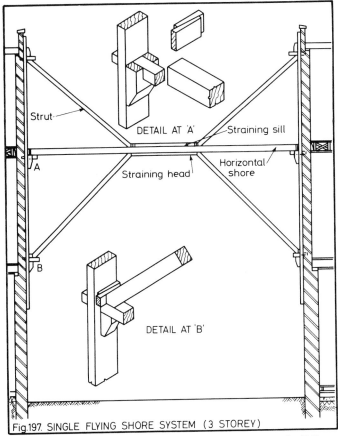

Fig.197. SINGLE FLYING SHORE SYSTEM (3 STOREY)

Strut
DETAIL AT 'A'
Straining sill
Horizontal shore
Straining head
A
B
DETAIL AT 'B'

conveniently level with each other as shown in Fig. 197. I see no reason why the normally horizontal shores should not be slightly inclined to accommodate these differences, although this seldom seems to be done, probably for the sake of appearance. The only other logical answer is to place the horizontal shore mid-way between the two levels of the opposing floors, if both buildings are of equal stability; otherwise, if one wall appears weaker either because it is bulging or lacks the anchorage of an adjoining floor as on the left of Fig. 199, then this should dictate the shore level. To reduce the risk of damage by shear to the other wall, a stiffener may be placed against the wall piece.

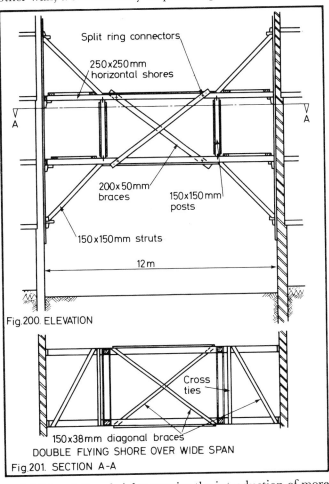

Split ring connectors
250x250mm horizontal shores
A
200x50mm braces
150x150mm posts
150x150mm struts
12 m
A
Fig.200. ELEVATION

Cross ties
150x38mm diagonal braces
DOUBLE FLYING SHORE OVER WIDE SPAN
Fig.201. SECTION A-A

Fig. 197 shows a shoring system to a three-storey building. It will be seen that a full length horizontal strut gives mutual support to the second floor, while the horizontal pressures at roof and first floor levels balance each other by means of raking struts thrusting against a straining sill and straining head respectively nailed to the continuous horizontal shore. Wall pieces, needles and cleats are used in the same way as for raking shores, the needle arrangement for the lower struts being inverted to take the downward thrust.

As struts are shorter and pressures more direct, smaller timbers may be used for the struts which may not be wide enough to notch around the needles as seen in detail at 'B'. Detail at 'A' enlarges on the cleats and needles carrying the horizontal shores and these, of course, are not subject to any thrust. All members are tightened by folding wedges of suitable sizes; their positions being such that if they become slack, they will not fall out.

The floors of opposing buildings will not always be

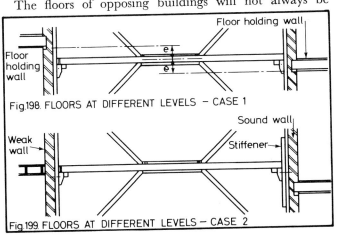

Floor holding wall
Floor holding wall
e
e
Fig.198. FLOORS AT DIFFERENT LEVELS – CASE 1

Sound wall
Weak wall
Stiffener
Fig.199. FLOORS AT DIFFERENT LEVELS – CASE 2

Additional floor heights require the introduction of more horizontal shores, as in Fig. 200, which has also been used to illustrate the precautions needed when the span between the buildings is great, say 12 m or more. There will be a greatly increased tendency to buckling of shores which will be accentuated by the need to form joints in their lengths because of the unavailability of long timbers. This buckling is best prevented by complete triangulation of the systems in pairs with diagonal braces and, where necessary, vertical or horizontal cross ties. The greatest risk of distortions is in the vertical plane with possible inequalities of pressure at different levels.

Where demolition is involved, the standing walls will have to be watched carefully and temporary struts inserted until space is available for the insertion of whole systems.

It is occasionally necessary to shore a high building off a lower one. In this case it is important to spread the load on the supporting structure by taking pressure from the top levels to more than one point in each system.

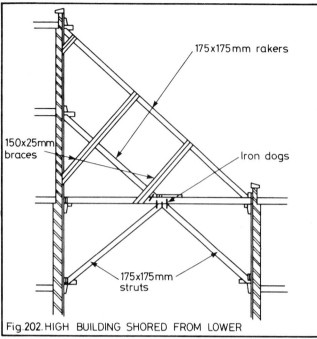

Fig.202. HIGH BUILDING SHORED FROM LOWER

(labels in figure: 175x175mm rakers; 150x25mm braces; Iron dogs; 175x175mm struts)

Fig. 202 illustrates this, where the load from the top floor level of the higher building is taken through by the alignment of the struts to that of the lower one, rather than accumulating pressures at the eaves level of the lower supporting building.

Shoring systems are normally vulnerable to the weather and settlement may be caused by a combination of shrinkage of the structural timbers as well as the possibly clay sub-soil in dry weather. Regular inspections should be made under these conditions.

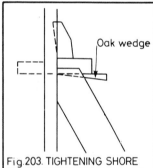

Fig.203. TIGHTENING SHORE

(label in figure: Oak wedge)

Raking shores are easily tightened by driving thin hardwood wedges under the needle as in Fig. 203. Existing folding wedges may have to be replaced by larger ones.

DEAD SHORES

As distinct from raking and flying shores, dead shores have to carry the weight of some part of the structure while that below is being cut away for repair or replacement, or for the formation of an opening to receive a steel or concrete lintel. In so far as the density, sizes and disposition of the elements are known, the dead loads from walls, roofs and floors can be calculated, while the maximum likely imposed loads from occupation, wind and snow can be estimated with reasonable safety.

The most common use of dead shores is in the formation of openings in stone or brick walls of dimensions suitable to receive shop windows or wide industrial doorways. Narrower openings up to 900 to 1000 mm width may sometimes be safely cut in brick walls which are in good condition, the adhesive effect of the mortar, plus the oversailing support of the bond, providing the necessary stability.

Above these widths, shoring is necessary, while the opening is being cut and the supporting beam inserted and pinned up.

The actual area of brick walling carried above the beam and in the interim supported by the shoring depends upon the degree to which the bonded walling on either side can be trusted to reach inwards. If the flanks on either side are narrow as in Fig. 204 or the bond indeterminate, it should be assumed that the whole of the wall will have to be carried, otherwise reduced loading as in Figs. 205 and 206 may be anticipated.

The essentials to safe dead shoring may be enumerated as follows:

1. The support given must be immovable and stable in itself and all bearing surfaces in compression must be fully under load before any cutting away is attempted.
2. All shores which support walls or any inflexible masonry construction must be taken direct to the ground or solid floor and must not be bridged by timber floors or ceilings.
3. Ground or foundation surfaces must be solid and compacted and if necessary levelled with weak concrete.
4. The shoring must be so positioned that it does not interfere with the operations of cutting out brickwork and the insertion and making good, of supporting beams.
5. The lateral stability of the structure must be main-

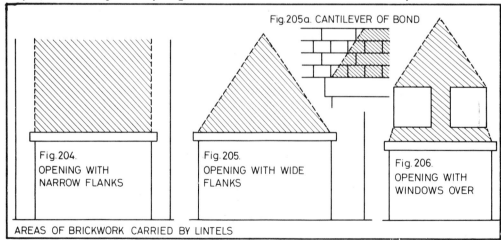

Fig.205a. CANTILEVER OF BOND

Fig.204. OPENING WITH NARROW FLANKS

Fig.205. OPENING WITH WIDE FLANKS

Fig.206. OPENING WITH WINDOWS OVER

AREAS OF BRICKWORK CARRIED BY LINTELS

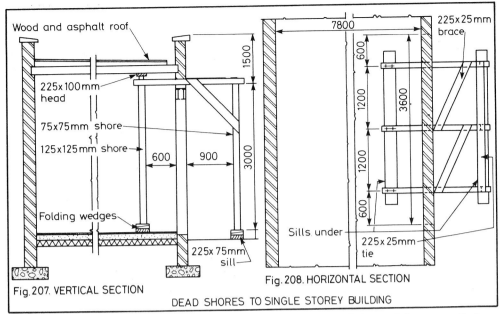

Fig. 207. VERTICAL SECTION

Fig. 208. HORIZONTAL SECTION

DEAD SHORES TO SINGLE STOREY BUILDING

tained, where necessary using flying or raking shores, as well as making the dead shore system stable in itself.
6. Where the wall carries floors or a roof, these must be supported independently where possible.

Figs. 207 and 208 illustrate the basic procedure in dead shoring. Support to the wall being cut away is provided at intervals by means of stout beams, termed 'needles', which are passed through apertures cut through the wall and are supported at each end by vertical props known as 'shores'. The needles have to be placed above the soffit position of the steel or concrete beams which eventually have to carry the wall as they cannot be removed until the beams have been bedded in and pinned up and the mortar hardened sufficiently. The spacing of the needles depends upon the type and condition of the wall and also upon the position of any windows as they must be confined to solid brickwork between the windows.

In Figs. 207 and 208 an opening is being formed in a single-storey building and the loading is comparatively light. The shores carrying the needles are 600 and 900 mm respectively from the inside and outside of the wall; thus giving working space for the bricklayer, particularly on the outside where most of the activity will take place.

The closer together the inner and outer shores are, the lower the bending moment and the smaller the needles can be. The feet of the shores are supported by continuous sills from which they may be tightened by means of folding wedges.

The wood and asphalt flat roof has its weight taken off the wall on to the inner shore by means of packing off the needles. Stability is given to the shoring system by a combination of ties and braces nailed to the needles and shores.

Calculation of timber sizes

This involves working out the sizes of the needles, treated as beams (see Fig. 209) and of the dead shores which are always in direct compression. When the building is of two or more storeys, a secondary support system given to floors and roof (Figs. 210 and 211) introduces further shores as well as heads placed under floors or ceilings which may be presumed to have uniformly distributed loads.

The first step in calculating member sizes is to obtain the total load carried by each member. Exact values are not necessary and would be difficult to obtain where the construction is concealed, so estimates should be on the conservative side. If the building is still in occupation, imposed loads must be added to the dead weight of the structure.

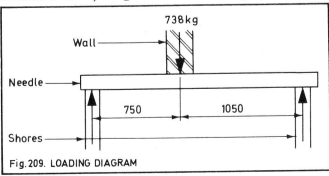

Fig. 209. LOADING DIAGRAM

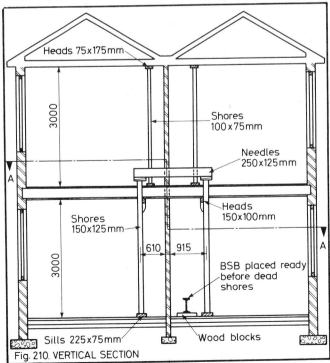

Fig. 210. VERTICAL SECTION

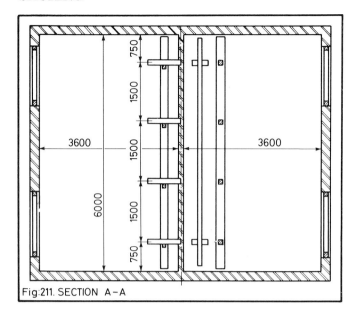

Fig.211. SECTION A–A

In any case it is safer to do this to allow for any material or equipment which may be deposited by the contractor.

The following list which is generous enough to cover traditional heavy construction should meet ordinary requirements.

LOADS ON STRUCTURES

Element	Dead load (kg/m³)
Pitched roofs (timber) with slates or tiles	137
Ceilings to the above	55
Timber flat roofs with asphalt or lead covering and plaster ceilings	137
150 mm concrete floors or roofs with ceilings	454
Domestic floors with ceilings (timber)	87
Timber floors to institutional buildings, offices, etc.; with ceilings	109
Half-brick walls	205
One-brick walls	410
One-and-a-half-brick walls	615
Two-brick walls	820

	Imposed Load
Pitched roofs	137
Flat roofs	274
Domestic floors	361
Institutional building floors	410
Floors of rooms of public assembly	820

Referring again to Figs. 207 and 208, the needles are not centrally loaded so the bending moment will have to be worked out. It is therefore necessary first to calculate the reaction at one of the supports, say the outer one. The load comes from the wall only; each needle takes a 1.2 m width of wall which stands 1.5 m above the proposed beam. The load on each needle therefore = $1.2 \times 1.5 \times 410$ kg = 738 kg = 7380 N.

$$\text{Outer reaction } Rr = 7380 \times \frac{750}{1800} = 3075\,\text{N}.$$

Therefore max. M = $3075 \times 1050 = 3228750$ N.mm

$$\text{Therefore } \frac{\sigma mbh^2}{6} = 3228750.$$

Using strength class SC3 timber we need to know
(1) Grade bending stress = $\sigma_{m,adm,\parallel} = 5.3$ N/mm²
(2) Grade compressive stress parallel to the grain $\sigma_{c,adm,\parallel} = 6.8$ N/mm²
(3) Grade compressive stress perpendicular to the grain $\sigma_{c,adm\perp} 1.7$ N/mm²....²

Then $h = \sqrt{\dfrac{6 \times 3228750}{5.3 \times 125}} = 171$ mm so a 125×175 mm

needle is needed. Note strength lost through high moisture content will be compensated by the higher strength capacity over short term loading. Next take the outer shore. This carries the end of the needle only, so that the load = 3075 N. A timber post can fail in two days when carrying a timber beam:
1. by crushing the cross grained fibres of the supported beam when it is stout, or
2. by buckling in itself when slender.

The approximate formula for reduction in effective compressional strength due to buckling is

$$\sigma_{c,e\parallel} = \sigma_{c,adm\parallel}\left(1 - \frac{H}{50b}\right)$$ when H equal post height and b =

least sectional dimension. Then ideally this should agree with $\sigma_{c,adm,\perp}$ (1.7 N/mm²). The best way is to take a trial value for b,

say 75 mm then $\sigma_{c,e,\parallel} = 6.8\left(1 - \dfrac{3000}{50 \times 75}\right) = 1.36$ N/mm².

Then the minimum safe sectional area of the post is $\dfrac{3075}{1.36}$

and $h = \dfrac{3075}{1.36 \times 75} = 30$ mm. But the minimum dimension

must be 75 mm so a 75×75 mm the nearest commercial sawn size.

Note: If $\sigma_{c,e,\parallel}$ came to more than 1.7 N/mm² then the actual bearing area of the needle would have to be checked. This would also of course vary if the thicknesses of both were not the same. Take the head over the inner shore. This carries the roof only, so with dead and imposed loads, load on each bay

$$= \frac{(137 + 274)718 \times 1.2}{2} = 1771\,\text{kg} = 17710\,\text{N}.$$

Taking $M = \dfrac{FL}{10}$ for a partly continuous beam with UDL.

$$\frac{\sigma bh^2}{6} = \frac{FL}{10}.$$

Taking $\sigma_m = 5.3$ N/mm², L = 1200, b = 225 mm and transposing and substituting,

$h = \dfrac{6 \times 17710 \times 1200}{10 \times 5.3 \times 225} = 103$ mm. So in practical terms a

100×225 mm head would be needed.

The imposed load could be omitted if the roof was not going to be occupied.

Taking the inner shore. Load from wall = $7380 - 3075 = 4305$ N. Load from floor (as for head) = 17710. Total load on shore = $4305 + 17710 = 22051$ N. Let $\sigma_{c,adm\parallel} = 6.8$ N/mm² and b = 125 mm then

$$\sigma_{c,e,\parallel} = 6.8\left(1 - \frac{3000}{50 \times 125}\right) = 3.5\,\text{N/mm}^2.$$

But the maximum for cross grained compression on the needle can only be $1.7 \, \text{N/mm}^2$. Therefore this value will have to be used.

The minimum cross section area of the

shore $= \dfrac{22015}{1.7} = 12950 \, \text{mm}^2$. Let $b = 125 \, \text{mm}$ then

$h = \dfrac{12950}{125} = 104$. Therefore the nearest commercial

size $= 125 \times 125 \, \text{mm}$.

Calculating dead shores for a two-storey building

Dead shores to an inner wall or a two-storey building when the lower part of the wall has to be removed are shown in Figs. 210 and 211. The shores are spaced at 1.5 m centres and the main shores are arranged to give a maximum of 915 mm from the wall. The wall and first floor have eventually to be supported, alternatively by one or two steel joists.

Fig. 212 shows one steel joist used to support the wall and floor cut off below the ceiling.

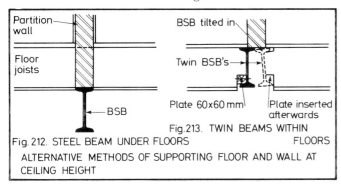

Fig. 212. STEEL BEAM UNDER FLOORS   Fig.213. TWIN BEAMS WITHIN FLOORS

ALTERNATIVE METHODS OF SUPPORTING FLOOR AND WALL AT CEILING HEIGHT

In Fig. 213, the steel joists are kept within the thickness of the wall giving an unobstructed ceiling. This is more expensive and difficult and involves cutting the ends of the joists to fit the steel beams. Two joists are necessary with a space between them so that, after one is fixed, the other can be tilted into position beside it.

The second supporting timber plate (not shown) is inserted as a final operation, if necessary in short lengths.

The steel beam or beams should be laid on the floor in readiness for hoisting before erecting the shores.

In order to take as much weight as possible off the needles, a supplementary system of shores and heads and sills are used to prop up the roof and first floor. Calculations involve the design of the heads under the roof and the shores which support them, the heads which support the first floor and the needles and shores carrying the wall.

In order to assess the actual loads carried by each element, it is simpler to take the whole of the roof and floor between external walls over a width of 1.5 m (spacing of all needles and shores) and then consider what proportion of this is carried by each element.

Starting at the roof, overall widths on slope $= 8.3 \, \text{m}$. Unit combined loads $= 274 \, \text{kg/m}^2$. Therefore total weight of 1.5 m strip $= 8.3 \times 1.5 \times 274 = 3411 \, \text{kg}$.

Upper ceiling over 1.5 m width $= 1.5 \times 7.2 \times 55 = 594 \, \text{kg}$.

Wall above needles $= 1.5 \times 3 \times 205 = 923 \, \text{kg}$ on each needle.

First floor load on 1.5 m width $= 7.2 \times 1.5 \times 448 = 4838 \, \text{kg}$. Separate elements are now as follows: Heads under roof, each will take 1/4 of ceiling span and 1/8 of roof span. Therefore the load on one bay

$$= \frac{594}{4} + \frac{3411}{8} = 5750 \, \text{kg} = 5750 \, \text{N}.$$

$$M = \frac{FL}{10} = \frac{\sigma bh}{6}.$$

Assuming $\sigma_m = 5.3$, and $b = 75$ and $F = 5850$.
Then by transposition and substitution

$$h = \sqrt{\frac{6 \times 5750 \times 1500}{10 \times 5.3 \times 75}}$$

$= 114 \, \text{mm}$ so a $75 \times 125 \, \text{mm}$ head will do.

Take the shores to the roof and assume least dimension cross section $= b = 75 \, \text{mm}$.

Then $\sigma_{cell} = 6.8\left(1 - \dfrac{3000}{50 \times 75}\right) = 1.36 \, \text{N/mm}^2$.

Load on shore $=$ load on head $= 5750 \, \text{N}$, so width of shore $= h$

$$= \frac{5750}{1.36 \times 75} = 56 \, \text{mm}$$ so a $75 \times 75 \, \text{mm}$

shore will do.

Take the needles. These support the wall and 1/4 of the roof.

$$\text{Total load F} = 923 + \frac{3411}{4} = 1776 \, \text{kg}.$$

Right hand reaction

$$R = \frac{610}{1525} \times 1776 = 710.4 \, \text{kg} = 7104 \, \text{N}.$$

Then max. $M = 7104 \times 915 = 6500160 \, \text{N/mm} = \dfrac{\sigma_m bh^2}{6}$.

Taking $\sigma = 5.3$ and $b = 125$. Then by transposition and substitution,

$$h = \sqrt{\frac{6 \times 6500160}{5.3 \times 125}} = 243 \, \text{mm},$$ so a 125×250 mm

needle will do.

Next take the lower head supporting the floor. Load from floor

$$= \frac{4838}{4} = 1210 \, \text{kg} = 12100 \, \text{N} \quad M = \frac{FL}{10} = \frac{\sigma_m bh^2}{6}.$$

Taking $b = 100, \sigma = 5.3, F = 12100$. Then by transposition and substitution,

$$h = \sqrt{\frac{6 \times 12100 \times 1500}{10 \times 5.3 \times 100}} = 143 \, \text{mm}$$

so a 150×100 head will do.

Next take the lower left hand shore. This will have to carry:

Part load from needle $= 1 \, \& \, 60 - 7104 = 10656 \, \text{N}$
Load from to shore $= 5750 \, \text{N}$
Load from floor $= 12100 \, \text{N}$
Total $= 28506 \, \text{N}$

SHORING

Assuming b (least cross section dimension = 125 mm and $\sigma_{c,adm,\|} = 6.8 \text{ N/mm}^2$ then

$$\sigma_{c,e\|} = 6.8\left(1 - \frac{3000}{50 \times 125}\right) = 3.536 \text{ N/mm}^2.$$

But 1.7 must be the maximum compression for cross grain so

$$h = \frac{\text{cross grain section}}{1.7b} = \frac{28506}{1.7 \times 125} = 135$$

so a 150×125 mm shore will be required.

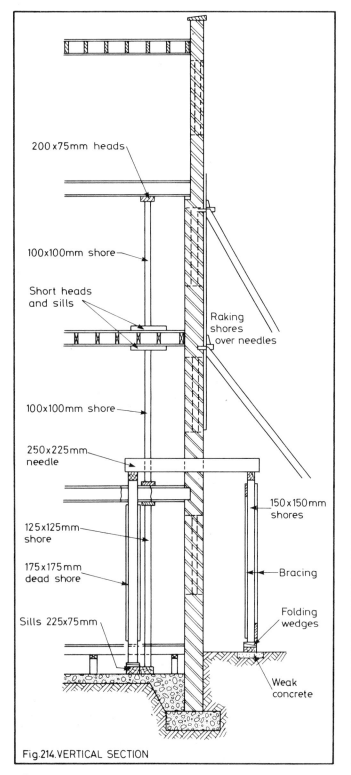

200x75mm heads

100x100mm shore

Short heads and sills

Raking shores over needles

100x100mm shore

250x225mm needle

150x150mm shores

125x125mm shore

175x175mm dead shore

Bracing

Folding wedges

Sills 225x75mm

Weak concrete

Fig.214.VERTICAL SECTION

Dead shoring for a four-storey building

The shoring system necessary to forming an opening in a four-storey building is shown in Figs. 214 and 215. Timber sizes are only approximate and have not been worked out. Although the same principles of construction arise as before, the following special details should be noted:

1. There is a hollow timber ground floor and shores have been taken through this to give a solid bearing on oversite concrete.
2. Due to the presence of windows on the upper floors, the shores and needles have to be spaced to come centre of the piers.
3. Intermediate floors, first and third, are supported separately to take their weight off the walls, but second floors and roof are carried by the return walls and do not need support, although propping has to embrace the second floor in order to reach the third.
4. The sill to the outer shores is evenly bedded on weak concrete under which the ground has been consolidated to ensure solid support.
5. Lateral stability is ensured: a. by tying the tops of the lower shores with longitudinal heads and bracing both ways, and b. by introducing raking shores at second- and third-floor levels.
6. Shores to the first floor are stouter as they have to carry the accumulated weight of the upper floors. Inner shores under needles are heavier as they are longer and would otherwise be more likely to buckle.

The presence of windows always presents an additional risk. These should be strutted as shown in Fig. 216. Where there is a fair amount of brickwork under the window, it is worthwhile retaining this by means of small secondary needles carried on beams or runners resting on the main

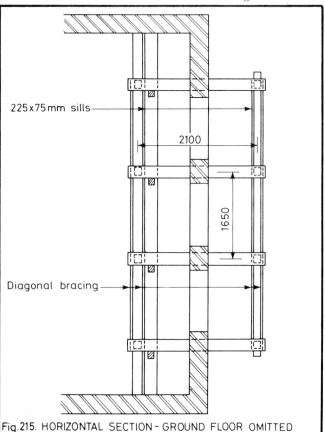

225x75mm sills

2100

1650

Diagonal bracing

Fig.215. HORIZONTAL SECTION – GROUND FLOOR OMITTED

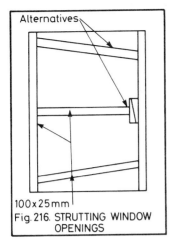

Fig.216. STRUTTING WINDOW OPENINGS

needles as in Figs. 217 and 218. These can be inserted at a later date as the brickwork beneath the window is being cut away.

It is important that all supported work should take a solid bearing before the supports come under load. For this reason, needles should be wedged up until they show slightly hollow on the top. Where the bearing surface of the wall is likely to be uneven, this can be adjusted by means of softwood packing wedged up as in Fig. 219. Alternatively the hole may have to be made good with cement mortar reinforced with a strip of expanded metal and given time to harden.

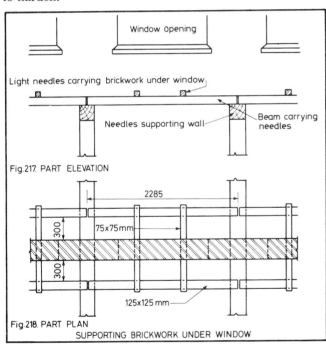

Fig.217. PART ELEVATION

Fig.218. PART PLAN
SUPPORTING BRICKWORK UNDER WINDOW

The work should not be started until everything required for the job, beams, etc. is on the job. All temporary supports given should start from the bottom and be fully tightened as they go. No cutting away should be done until all supports are in place.

The wall should be cut away at each end of the proposed opening, the wall made good, piers formed, and bearing provided for the beam before cutting away the rest of the wall.

Fig. 220 shows a method of getting the supporting BSB into place. It is hoisted by block and tackle and then supported by bearers resting on cleats already nailed to

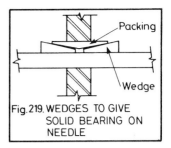

Fig.219. WEDGES TO GIVE SOLID BEARING ON NEEDLE

convenient shores. A couple of fillets nailed to the top of each bearer provide a groove into which a lever may be inserted to work the beam into position. Striking of shoring should be done very carefully avoiding any risk of shock, in the reverse order to erection, at least a week being allowed for the work to settle and harden.

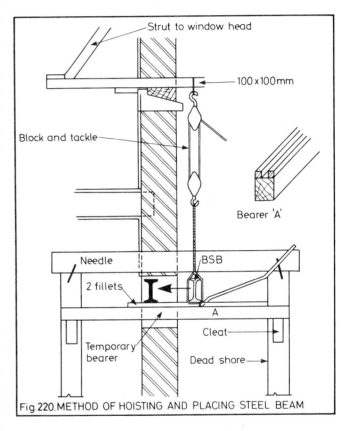

Fig.220. METHOD OF HOISTING AND PLACING STEEL BEAM

Special shoring problems

There is always the likelihood of meeting special shoring problems to which the ordinary routine cannot be applied. One such is illustrated in Figs. 221 to 224. This relates to the requirement to repair or replace a column in an arcade to a church aisle.

The problems here are to take the weight of the capital or top part of the column, to relieve it as much as possible of the load from the supported spandrel, and to give complete stability to this very vulnerable part of the structure.

Fig. 224 shows a timber yoke cut to the shape of the column shaft. This is tightened around the shaft by means of bolts one way and by wedges driven against a tenon the other (note the clearances for tightening). This should also grip under any projecting mould to the capital. The spandrel is supported by curved ribs supported by a system of struts and ties which also carry the weight of the yoke.

57

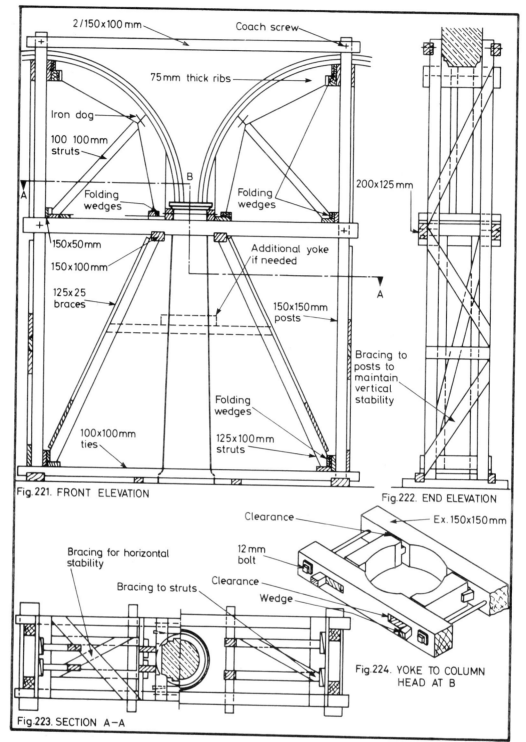

Fig. 221. FRONT ELEVATION

Fig. 222. END ELEVATION

Fig. 223. SECTION A—A

Fig. 224. YOKE TO COLUMN HEAD AT B

Vertical posts are erected either side of the arcade, tied and braced through the wall thickness and tied together at the top longitudinally with two runners. The posts rest on sills which should preferably be Rawlbolted to the floor. The load carried by the yoke is taken on two transoms which rest on headtrees strutted back to the sills.

The arches are supported at the crown by the posts, at the springing by the struts through the transom and headtrees and mid-way by struts taken back to the ends of the transoms. To give lateral stability, everything is braced and tied where possible, the transom, the posts, and the struts. Erection is best carried out as follows:

1. Erect the vertical posts, and sills and the upper and lower ties and brace as shown in Figs. 221, 222 and 223.
2. Bolt and wedge the yoke around the column, attach the transom headtrees and struts and wedge up tight with folding wedges and brace the struts.
3. Cut hardboard templets roughly to fit the arch curves, scribe to fit with compasses, cut and use to mark out the ribs.
4. Cut and fit the ribs, strut into position and tighten at all points by means of folding wedges. If the top part of the column is to be retained, this may be supported by an additional yoke.

CHAPTER 6

Roofing

In deciding the slope of a roof, the need to make it weatherproof is the first consideration. If it is covered with a continuous sheet material, lead, copper, aluminium, etc.; then, provided that the joints are sealed or lifted clear of the roof surface, a slope of 1:80 is sufficient to disperse standing water over minor irregularities and may be regarded as satisfactory. When, however, the roof is covered with overlapping small units such as slates, tiles or corrugated sheeting, then the slope or pitch must be sufficient to take the water away with some rapidity before, assisted by wind and capillary action, it can seep back into the building below.

It follows that the shorter the lap of the tile the greater the vulnerability of the joint, and as the lap must be practically related to the tile length, short units such as plain tiles must be laid at a greater pitch than say large corrugated sheeting, which is seldom less than 1.5 m long.

SIMPLE PITCHED DOMESTIC ROOFS

The minimum pitch needed for a specific roof covering may vary with site conditions, degree of exposure, etc., while in most cases some adjustment may be made in the lap. The values below, therefore, may only be taken as average:

Covering	Deg. pitch
Large corrugated sheeting	15
Large slates	$22\frac{1}{2}$
Small slates and single lap tiles	35
Plain tiles	40
Thatch	45

It will be appreciated on consideration that the pitch of the tile is less than that of the rafters (Fig. 225).

In deciding on the actual roof structure, it is common practice, as recommended in most textbooks, to relate this closely to the roof span. Thus a lean-to roof is given a maximum span of 2.4 m, a couple roof 3.6 m, a collar beam roof as 5.4 m and so on.

The better way, in my opinion, is to start with the basic units, the rafters which support the roof covering and the joists carrying the ceiling, both of which act as beams, then, having decided on economic sections and the spans over which timbers of these sections will carry their loads, use these as unit lengths to be given end supports over the required span.

The principles involved are illustrated in Figs. 226 to 234. Thus, assuming rafters of a given section and spacing will carry a calculated dead load plus imposed load over a span 'S' while the equivalent span for ceilings is 'C', then a single span will permit the use of a 'lean-to roof (Fig. 226). Twice this span will need a couple roof (or if a ceiling is required a couple close as in Fig. 227).

The fink truss in Fig. 228 combines four rafter spacings with three ceiling joist spans, while the built-up timber roof

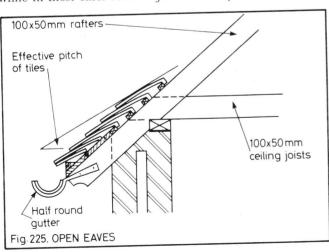

100x50mm rafters

Effective pitch of tiles

100x50mm ceiling joists

Half round gutter

Fig. 225. OPEN EAVES

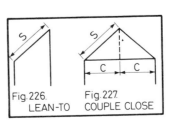

Fig. 226. LEAN-TO

Fig. 227. COUPLE CLOSE

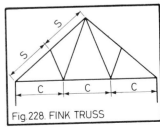

Fig. 228. FINK TRUSS

ROOFING

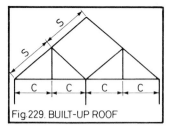

Fig.229. BUILT-UP ROOF

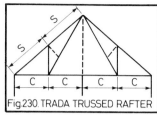

Fig.230. TRADA TRUSSED RAFTER

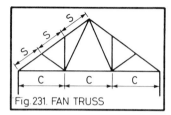

Fig.231. FAN TRUSS

Fig. 229 and the TRADA roof Fig. 230 give four bays of each. The fan truss roof Fig. 231 has six bays of rafters and three of ceiling joists.

The industrial-type roof truss in Fig. 232 shows the same principles applied over a wide span. Fig. 233 shows a badly designed roof truss in that the pitch of the struts is very flat. This would result in

1. Their being subject to very high stress, and
2. A minimum amount of strain would allow a large deflection.

Fig. 234 shows a collar tie roof. This again is inefficient as the ends of the rafters are subject to a high cantilever loading, but may be used where head room is required above wall plate level.

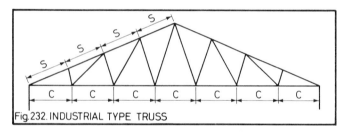

Fig.232. INDUSTRIAL TYPE TRUSS

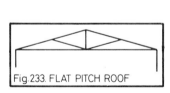

Fig.233. FLAT PITCH ROOF

Fig.234. COLLAR TIE ROOF

If 'S' is given an average value of, say, 2.4 m, then this may give maximum spans for various constructions; but by adjusting the timber sections of rafters and joists in relation to the loads, 'S' and 'C' can both be modified to give a reasonable amount of flexibility.

Common details of construction

Domestic pitched roofs are designed to take small roofing units, clay or concrete tiles or slates. These are carried by horizontal battens nailed at close intervals to common rafters. The spacing of the rafters governs the section of the tiling battens, but it may have to coincide with that of the ceiling joists to suit the safe span of plaster boards for the ceiling.

Alternative spacings are commonly 400 mm, 450 mm or 600 mm. The wider the spacing, the greater the load on the ceiling joists and rafters, so that this forms another feature by which member spans may be modified.

Intermediate support may be given to rafters by purlins carried on cross walls, or on trusses, or strutted down to load-bearing partitions, or purlins may be omitted and each pair of rafters can be made up with the ceiling joist into light trussed frames.

Lean-to roofs

A lean-to roof is the simplest type of pitched roof. Common rafters are birds-mouthed over a wall plate at the feet and carried by a wall plate or wall piece at the head as shown in Figs. 235 to 237.

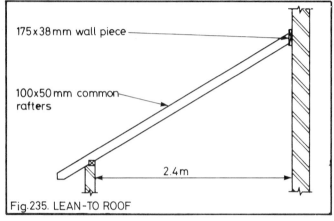

175 x 38 mm wall piece

100 x 50 mm common rafters

2.4 m

Fig.235. LEAN-TO ROOF

It is commonly believed that a raking member, when loaded, must exert a horizontal thrust. This is not necessarily so; and provided that the top and bottom of a raker are adequately supported with level joints, there is no overturning force whatever from a vertical load.

This can easily be proven with three pieces of wood, carefully cut as in Fig. 238, and a brick. The set-up will still stand after the load has been applied. If the lean-to in Fig. 235 is inadequately nailed, it will, of course, slide down the main wall and overturn the lower wall.

If, however, it is birds-mouthed over the top of a wall plate carried by a corbel, as in Figs. 236 or 237, it will carry its maximum load even before nailing. The only criterion therefore is the strength of each rafter to resist bending.

Fig. 239 shows a section through a lean-to roof to form an extension, say, to a main building with a ceiling above wall plate level. If a ceiling joist is nailed to each rafter as

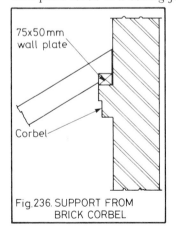

75 x 50 mm wall plate

Corbel

Fig.236. SUPPORT FROM BRICK CORBEL

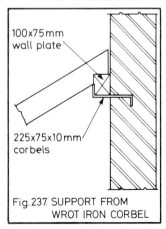

100 x 75 mm wall plate

225 x 75 x 10 mm corbels

Fig.237. SUPPORT FROM WROT IRON CORBEL

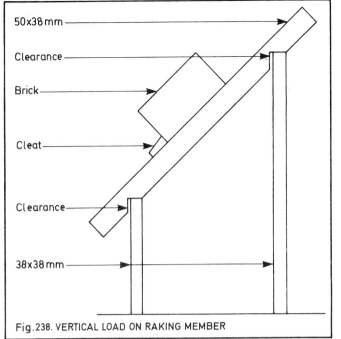

Fig.238. VERTICAL LOAD ON RAKING MEMBER

Labels: 50x38 mm, Clearance, Brick, Cleat, Clearance, 38x38 mm

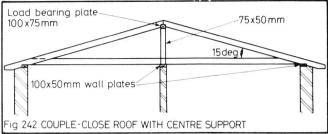

Fig. 242. COUPLE-CLOSE ROOF WITH CENTRE SUPPORT

Labels: Load bearing plate 100 x 75mm, 75x50mm, 15deg, 100x50mm wall plates

shown, then this will support the rafter and reduce its bending moment, thus permitting a smaller section to be used. The effect of the load will now, however, by acting against the ceiling joist at point 'P', result in a thrust along the roof line and tend to pull the wall plate away at the top and overturn the lower wall.

If, therefore, one wrought iron corbel is built in to each rafter, the rafter securely nailed to each plate and the ceiling joist halved over and well nailed to the rafter, this construction will prove satisfactory and will actually anchor the lower wall in position.

Fig. 242 shows a roof over a small building with two bays of, say, 2.4 m each formed by a central partition such as would be required for a double garage to two adjoining properties. If a couple close roof were used, half the weight of the roof would be on each external wall. Assuming that the load carried by one rafter = 1657 N, then the frame stresses, excluding bending, are represented by the space diagram Fig. 243 where 'AC' represents the known vertical reaction and 'CB' and 'BA' the lines of stress in the tie and rafter.

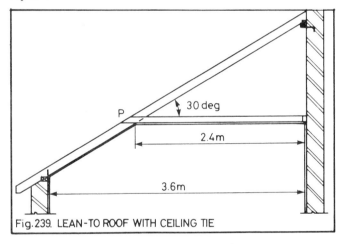

Fig.239. LEAN-TO ROOF WITH CEILING TIE

Labels: P, 30 deg, 2.4m, 3.6m

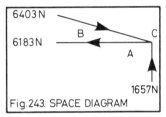

Fig. 243. SPACE DIAGRAM

Labels: 6403N, 6183N, B, C, A, 1657N

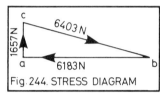

Fig. 244. STRESS DIAGRAM

Labels: c, 6403N, 1657N, a, 6183N, b

From the stress diagram in Fig. 244, it will be seen that both rafter and tie are stressed far in excess of the actual load. If, however, a load-bearing plate is strutted off the central wall as shown in Fig. 242 with struts, say at 1 m centres, half the weight of the roof will be carried by the central wall, the rafters will be subject to direct bending only and the horizontal tie will be virtually redundant.

Collar tie roof construction is generally limited to buildings in which for economy reasons the ceiling level is kept above that of the wall plates.

In the past, various types of cut joints, such as halving, dovetail halving or birdsmouth notching, have been used to form the joint between collar tie and rafter. The serious disadvantage of this lies in the fact that the joint occurs at

If the lines of force acting at 'P' are represented by the space diagram in Fig. 240 in which half the dead load plus imposed load carried by one rafter are represented by the force 'AB' as 1450 N then the pull on the rafter = line 'BC' and the thrust on the ceiling joist = CA. A stress diagram drawn to scale (triangle of forces Fig. 241) will give the tension on the top of the rafter as 2900 N and the horizontal pull on the wall plate and thrust on the ceiling joist as 2511 N.

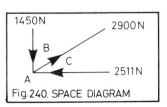

Fig.240. SPACE DIAGRAM

Labels: 1450N, 2900N, B, C, A, 2511N

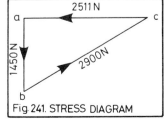

Fig.241. STRESS DIAGRAM

Labels: 2511N, a, c, 1450N, 2900N, b

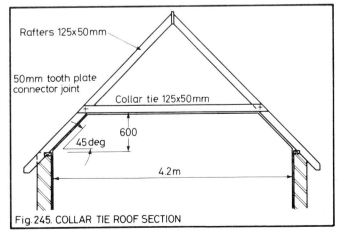

Fig.245. COLLAR TIE ROOF SECTION

Labels: Rafters 125x50mm, 50mm tooth plate connector joint, Collar tie 125x50mm, 600, 45deg, 4.2m

the point of maximum stress of the rafter (if the tie is kept appreciably high and the resistance of the wall to over-turning is ignored). Thus the sectional area only of the timber left in the rafter after cutting the joint must be considered for its design strength.

In the example given in Fig. 245, each joint has been made with tooth plate connectors. This does not reduce the effective section as the bolt hole comes in the neutral fibres. The dead and imposed load on one rafter will also be the reaction at its support. The cantilevering of the rafter below the tie will be the component of the load acting at right-angles to the rafter.

Taking the left-hand side (Fig. 245), clockwise moments will be the cantilever effect of the reaction at wall plate level as a point load. Anti-clockwise moments will be the uniformly distributed load below the tie and the maximum bending moment will be the difference.

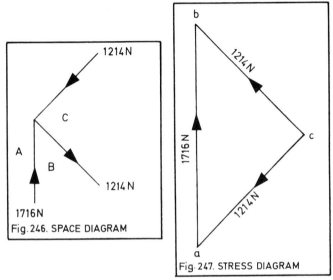

Fig. 246. SPACE DIAGRAM

Fig. 247. STRESS DIAGRAM

Assuming a pantile roof and taking accepted values for dead and imposed loads with rafters spaced at 450 mm centres, the weight carried by one rafter and reaction at support = 1716 N. Reaction at 90 deg. to rafter (Figs. 246 and 247) = 1214 N. Length of cantilever = 600×1.414 = 848 mm.

$$\text{Load below tie} = \frac{600}{2100} \times 1214 = 347 \text{ N}.$$

$$\text{Then CWM} = 1214 \times 848 = 1029472 \text{ Nmm}$$

$$\text{ACWM} = 347 \times \frac{848}{2} = 147128 \text{ N/mm}$$

$$\text{Maximum M} = 882344 \text{ N/mm}$$

$$M = \sigma Z = \frac{\sigma bh^2}{6} \quad \text{Let } \sigma = 7.3 \text{ N/mm}^2 \text{ and try } 125 \times 50 \text{ mm}$$

$$\text{rafter then } \sigma Z = \frac{7.3 \times 50 \times 125^2}{6} = 950521 \text{ N/mm}$$

which is in excess of what is required for bending. The rafter is also subject to longitudinal compression, but as it will be adequately tied by battens, etc., the full value for compression, say $\sigma_c = 8.0 \text{ N/mm}^2$ then safe compression = $8 \times 125 \times 50 = 50\,000$ N.

The sum of the fractions

$$\frac{\text{Actual } \sigma Z}{\text{max } \sigma Z} + \frac{\text{Actual Com.}}{\text{Max. Com.}} \leqslant 1$$

$$\frac{882344}{950521} + \frac{1214}{50\,000} = 0.928 + 0.024 = 0.952$$

So a 125 by 50 mm joist should be satisfactory.

In these calculations the weight of the ceiling has been ignored, as has also been the lateral resistance of the walls. It is reasonable to presume they will offset each other.

The collar tie has not been calculated. This would be subject to bending and tension and values for the connectored joint may be obtained from BS 5268 part 2. There would be tension parallel to the grain for the tie and at 45 deg. to the grain for the rafter.

Built-up roofs

Built-up roofs (Fig. 248) are common to normal small dwellings where a load-bearing partition wall is at or near the centre of the building or cross walls are available at convenient intervals to provide support.

Rafters are birdsmouthed over the wall plate and cut against a 25 mm thick ridge. Ceiling joists also act as ties and if in two lengths are lapped or cleated together at the joints over the central wall. Struts may be birdsmouthed over a plate nailed to the ceiling joists and are better notched around the purlins. They should be stiff enough to resist buckling, particularly if they are long.

The ceiling joists will need support at mid-span between the walls and this is most easily done with binders suspended

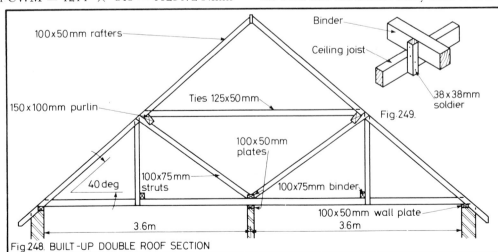

Fig. 249.

Fig. 248. BUILT-UP DOUBLE ROOF SECTION

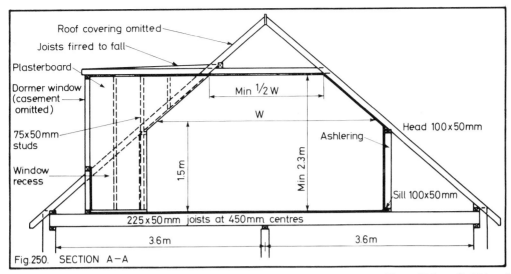

Fig.250. SECTION A–A

by hangers from rafters adjoining the purlins. The hangers should lap and be nailed to ceiling joists where they occur; while intermediate ceiling joists should be secured by soldiers pre-drilled to avoid splitting and nailed to binder and joist (Fig. 249).

The upper horizontal ties may be considered redundant, but they do form a couple close with the top half of the rafters, help to hold the purlins and transfer some of the raking thrust from rafters on to the struts.

The ties, struts and hangers are all equally spaced and come together at intervals of about 1.5 to 2 m. Increasing the spacing of these will put more load on the purlins and binders and then on to struts and hangers, the size of these being related to the above spacing.

At the present time, the formation of attics within the roof in old buildings and new is a popular practice in domestic buildings. The problem with these is that in order to provide a clear room space little is left for internal strutting. Some advantage may be taken of internal cross walls or timber framing within transverse partitions in the attic to carry the floor; but if there are load-bearing partitions to the floor below, these can be used to support floor joists to the attic, these being made stout enough to carry the greater part of the weight of the roof through the vertical ashlaring as well as the floor.

This is shown in Figs. 250 and 251 which also show structural details for a dormer to provide light. To comply with the Building Regulations, the ceiling height must be at least 2.3 m at which level the ceiling area must be at least half of that taken on a horizontal plane 1.5 m off the floor.

It is the present-day custom to provide light partitions within domestic buildings. As these are seldom able to carry any load, the whole of the roof and ceiling must be self-supporting. One of the earlier methods of providing for this is by the TRADA trussed rafter system as shown in Figs. 252 to 260. In effect, every fourth pair of rafters is built into a truss which also carries the other rafters and ceiling joists between by means of purlins, binders, wall plate, and ridge, an outline section through these being given in Fig. 253.

Figs. 255 to 260 are pictorial sketches of the joints between members which are assembled with 3 mm thick washers and 12 mm bolts carrying tooth plate connectors sandwiched between all contact surfaces. The teeth must be fully embedded into the timber on both sides and special equipment is available to ensure this.

It will be seen from the plan Fig. 254 studied in relation to the pictorial sketches, that there is an element of twist in the construction which varies slightly with different sizes

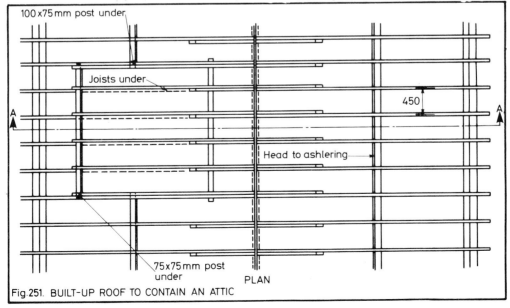

Fig.251. BUILT-UP ROOF TO CONTAIN AN ATTIC

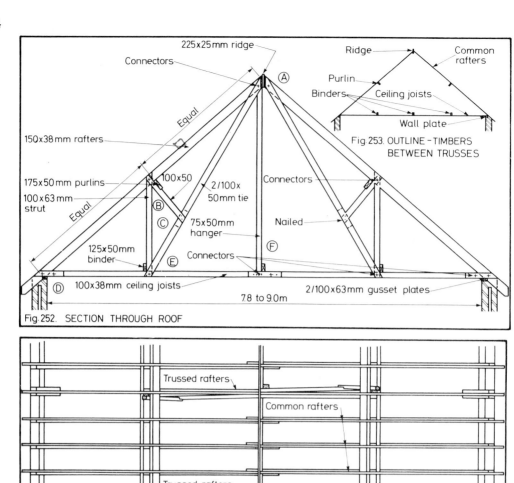

Fig. 252. SECTION THROUGH ROOF

(labels: 225×25mm ridge; Connectors; Equal; 150×38mm rafters; 175×50mm purlins; 100×63mm strut; 100×50; 2/100×50mm tie; Equal; 125×50mm binder; 100×38mm ceiling joists; 75×50mm hanger; Connectors; Nailed; Connectors; 7.8 to 9.0m; 2/100×63mm gusset plates; A, B, C, D, E, F)

Fig. 253. OUTLINE – TIMBERS BETWEEN TRUSSES

(labels: Ridge; Common rafters; Purlin; Binders; Ceiling joists; Wall plate)

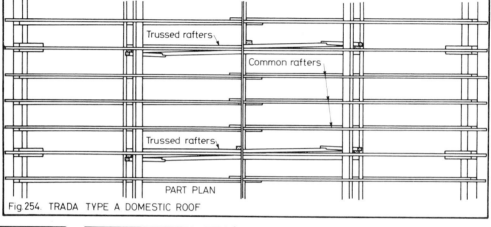

PART PLAN

Fig. 254. TRADA TYPE A DOMESTIC ROOF

(labels: Trussed rafters; Common rafters; Trussed rafters)

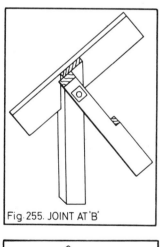

Fig. 255. JOINT AT 'B'

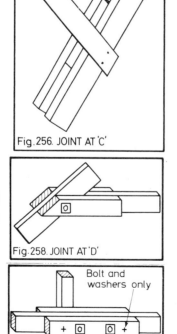

Fig. 256. JOINT AT 'C'

Fig. 257. JOINT AT 'A'

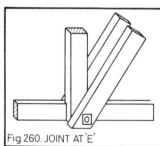

Fig. 258. JOINT AT 'D'

Fig. 259. JOINT AT 'F'

(label: Bolt and washers only)

Fig. 260. JOINT AT 'E'

of truss. The twist is not, however, sufficient to affect stability or make construction difficult.

The trusses are made up in halves, both the same hand, bolted together on site. The members into which the purlins are notched merely act as brackets to carry the purlins and are therefore only nailed at the lower ends. Also, if the ceiling joists can be supported by a centre wall, the hanger and centre binder become unnecessary. These items are shown dotted in Fig. 230.

Eaves finishes have not varied much over the years. Fig. 225 shows an open eaves construction. A triangular fillet from 38 to 50 mm deep is necessary to replace the discontinuing tile and also tilt the bottom edges up and close the joints. The feet of the rafters and face of the board soffit must be given a planed finish. The brickwork, continued between the rafters and known as beam filling helps to insulate the roof by cutting out draughts.

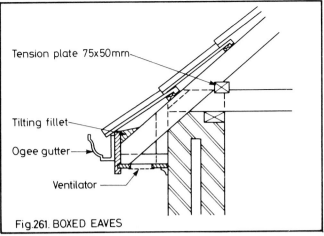

Fig. 261. BOXED EAVES

Fig. 261 shows boxed or closed eaves. The fascia stands below the soffit to act as a drip. It is necessary to ventilate the loft area to prevent condensation on the underside of the roof felt. This is done through the soffit at the eaves. It may be done either with continuous ventilation, the full length of the eaves or by a series of wider openings. In either case the effective ventilation area must be at least equivalent to a continuous opening in the soffit 10 mm wide in the case of roofs of a pitch greater than 15 deg. or 25 mm when the pitch is less than 15 deg. The openings should be protected against flies with a suitable plastic or metal gauze Mono pitched roofs should also be ventilated at the ridge. In normal roof construction there is considerable compression in the foot of the rafter and tension in the ceiling joist. A notched tension plate has been shown to accommodate these stresses.

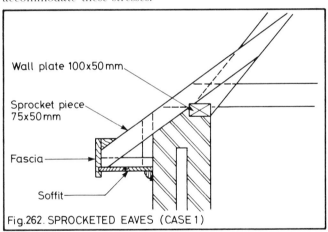

Fig. 262. SPROCKETED EAVES (CASE 1)

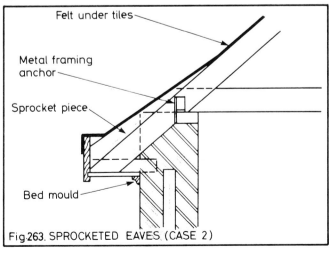

Fig. 263. SPROCKETED EAVES (CASE 2)

In Fig. 262, the pitch of the roof has been lowered at the foot by sprocket pieces nailed to the sides of the rafters. This is sometimes done for appearances' sake to make the roof look less overpowering, but it also has the effect of checking the flow of the rainwater on a steep pitch where it might otherwise shoot out over the top of the gutter.

Fig. 263 is an alternative construction where the sprocket is nailed to the top of the rafter. It should be noted that this does not replace the tilting fillet. Also shown is a three-sided box plate metal framing anchor. Clout nails—32 mm No. 11—are driven through holes in the plate into the rafter and two sides of the wall plate. The roofing felt which should be used on all domestic pitched roofs except under shingles has also been shown here to indicate its finish.

A popular opinion is that it should turn down into the gutter. I disagree with this on two points.

1. There is no satisfactory fixing to the edge of the felt, and
2. The rain can blow into the gutter and drive it in under the felt and over the vulnerable unpainted top edge of the fascia.

If the felt is nailed as shown, any seepage (which is all that there should be) will run down the vertical face of the fascia.

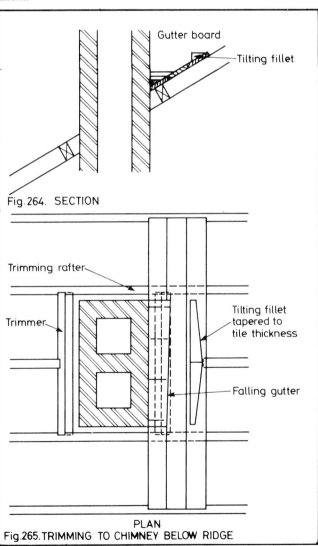

Fig. 264. SECTION

PLAN
Fig. 265. TRIMMING TO CHIMNEY BELOW RIDGE

65

ROOFING

Where the chimney passes through the roof the rafters need to be trimmed as shown in Figs. 264 and 265. To conform with the Building Regulations, no structural timbers should be within less than 40 mm of a half-brick flue. When the chimney passes through the apex, the ridge must be cut back and may be supported by brick corbels built out from the chimney (within the limits of the above regulation).

The drawings show the trimming to a chimney below the apex. A gutter must be formed behind the chimney to take the necessary plumber's work. The gutter should have a fall one way; a tilting fillet also is necessary. This should be tapered at the ends so that the outer tiles can line up with the main roof tiles.

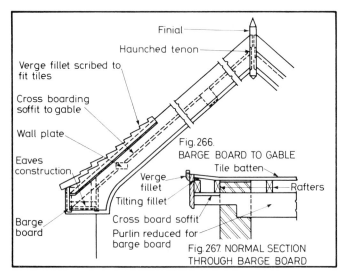

Fig. 266. BARGE BOARD TO GABLE

Fig. 267. NORMAL SECTION THROUGH BARGE BOARD

Finish to gables can be either in timber or the roofing timbers can finish against the inside of the gable wall with battens and tiles carried over.

Figs. 266 and 267 show the gable end finished with barge boards. Purlins and wall plates are carried through the wall to give the necessary overhang to support a pair of rafters on each pitch as in Fig. 267. A barge board 38 mm thick is nailed to the outer rafters and a soffit of short boards or some other suitable sheet material nailed up under the rafters. The barge board must extend and widen at the bottom to cover the eaves construction.

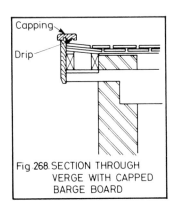

Fig. 268. SECTION THROUGH VERGE WITH CAPPED BARGE BOARD

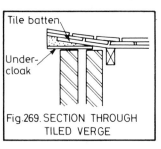

Fig. 269. SECTION THROUGH TILED VERGE

The tiling battens as shown should be tilted to throw the water back on to the roof. The verge may finish with a scribed fillet as in Fig. 266, or the barge board may finish above the tiles with a wooden capping as in Fig. 268. The ends of the barge board are tenoned at the top into a finial

which is also mortised to receive a tenon on the end of the ridge. The mortises should be designed to miss one another.

An alternative verge detail is as shown in Fig. 269 but this is entirely the concern of the tiler.

CONSTRUCTING TRUSSED RAFTERS

The need for speed in domestic building has led to the maximum amount of prefabrication of structural units with the attendant reduction in site labour, particularly with regard to roofs, the rapid completion of which, of course, provides early protection against inclement weather. One development from this need, has been the introduction of the single trussed common rafter, every pair of rafters being formed into a complete truss capable of supporting its own portion of roof covering and ceiling over the whole span of the building. Although there may be four or six times as many internal struts and ties (web members) as are needed for conventional built-up roofs, or with the TRADA type trussed rafters, the system is economically viable.

Economy arises from the following factors:

1. The trusses are made to a simple design with butt joints strengthened by (what are in effect) metal, or plywood gussets and can be rapidly assembled with the equipment relevant to the particular type.
2. As members are of uniform thickness and assembled in a flat plane they are easily stacked both for transportation and on site.
3. The excess of web members is more than compensated by the absence of ridge and purlins.
4. The trusses are light and the roof can be rapidly assembled, the trusses being merely lifted on to the wall plates nailed in position, braced, and strutted upright.
5. The units are widely spaced, say, at 600 mm crs and although stouter battens will be needed there will still be an overall saving in timber and labour.

Design of trusses may vary according to the span and the type of gusset (or its equivalent) used. This may be either:

1. Plywood nailed to connected members.
2. Plywood glued to members.
3. Metal plates nailed over the joints through predrilled holes or

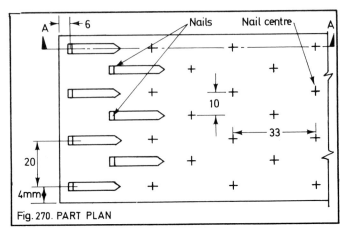

Fig. 270. PART PLAN

66

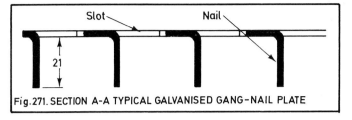

Fig.271. SECTION A-A TYPICAL GALVANISED GANG-NAIL PLATE

4. Proprietary Gang-nails which are metal plates with the nails integral with the plates, being pressed out of them at right angles as shown in Figs. 270 and 271.

Method 2. can be used with nails spaced so that each nail holds about 5000 mm² of glued area (71 mm spacing); but the method is, in my opinion, risky even with a gap-filling glue, as the impact of the hammer from which the contact surfaces may slightly recover tends to starve the joint of adhesive. This should be a factory operation using jigs and special clamps. Gang-nails can only be used with special equipment which includes a jig for holding members in place and high power presses to force the teeth of the plates into both sides of the joint.

The only methods likely to be used by the site carpenter would be that with nailed plywood gussets or drilled metal plates, particularly the first of these which will now be considered.

The information now required is:

1. The type of truss to be made (fink or fan type) as previously outlined.
2. The grade of timber and its moisture content.
3. The sizes of rafters, ceiling ties and internal struts and ties (webs).
4. The type, thickness and grade of plywood used.
5. The nailing pattern, i.e. number and spacing of nails, and from that, the overall size of each pair of gussets, also the type and size of nail.

The Building Regulations 1985 require that all timber structures should conform to the relevant parts of BS 5268, Part 2 or 3 and anyone concerned with the design of these trusses should be familiar with these codes of practice. Building Regulations 1985 give tables for roofing timbers with inter-related loadings, sizes, spans and spacings of rafters, ceiling ties, purlins and ridges of normal built up roofs. However, in this type of construction structural timbers are concerned largely with bending stresses only, points loads being taken

through struts to the walls. In trussed rafter construction, the whole of the roof load is transmitted via rafters and ceiling ties to external walls. Rafters are therefore subject to both bending and axial compression whilst ceiling ties must resist bending and tension. For information on timber sizes related to these conditions reference should therefore be made to BS 5268 part 3. This code of practice recommends three alternative methods of dealing with trussed rafter design. They are, briefly,

(1) Structural analysis by the use of engineering calculations
(2) Prototype testing of full sized constructions
(3) partial design by taking sizes of rafters ceiling ties and webs from the tables given in the code and then designing or otherwise obtaining information on the joints to ensure that they are of adequate strength.

To this end several associations involved in promoting the use of materials with which member firms are concerned, issue design brochures on the site manufacture of trussed rafters using nailed plywood or metal gussetts or plates. Sizes of timber given tend to be in excess, but information on nailed gussets may be used with some interpolation on intermediate sizes. One example is a booklet issued by The Finnish Plywood Development Association (FPDA) entitled "The design of roof structures using Finnish Plywood".

Table 5 in BS 5268 part 3 gives lengths of struts related to breadth and depth and tables 13 and 14 sizes of rafters and ceiling ties respectively related to span and pitch for three different grades of timber. In each case also values are related to two different thicknesses of PAR timber, i.e. 35 mm and 47 mm thick and within these limits related spans and pitches are given for all the depths of commercially available sizes. There is therefore no point in interpolating. For intermediate spans, the next size over is taken.

Taking as an example a roof span 10 m with 30 deg. pitch. It is proposed to take the rafter sizes from the BS tables and assemble with plywood gussets as given in the FPDA design sheets. The thickness of timbers is taken as 47 mm to agree with the sawn 50 mm dimension in the design sheet.

From the tables in the code using SC4 timber, rafters will need to be 47 × 120 mm, ceiling ties 47 × 97 mm and webs 47 × 72 mm. Referring to the design sheets of FPDA these were all for minimum pitch of 17½ deg. The flatter the pitch the more strain on the joints, so for a 30 deg. pitch this is a factor of safety. If the gussett surface dimensions are taken as for a 11 mm span this is another factor of safety. The design

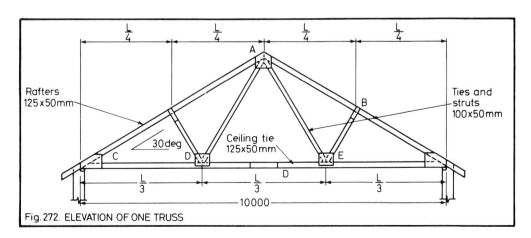

Fig. 272. ELEVATION OF ONE TRUSS

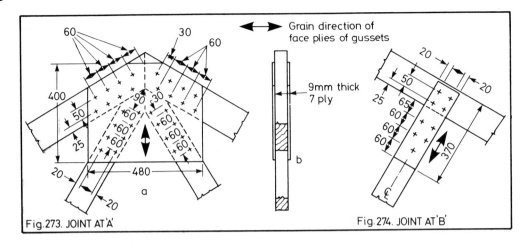

Fig.273. JOINT AT 'A'

Fig.274. JOINT AT 'B'

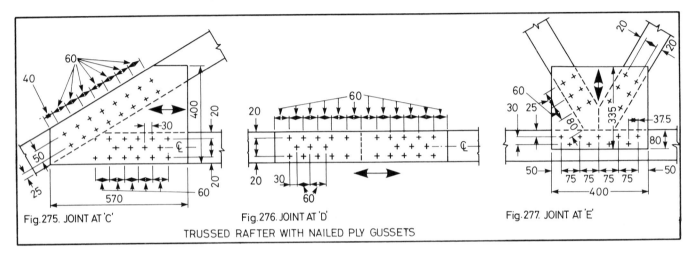

Fig.275. JOINT AT 'C'

Fig.276. JOINT AT 'D'

Fig.277. JOINT AT 'E'

TRUSSED RAFTER WITH NAILED PLY GUSSETS

sheets give 6.5 mm thick 5 ply gussetts with 9 swg 32 mm long square twisted nails for the 9 m truss and 9 swg 50 mm square twisted nails with 9 mm thick 7 ply gussetts for the 11 m truss. It would seem to be reasonable to use the thinner ply with the shorter nails for the 10 m span trussed rafter. Fig. 272 is the elevation and Figs. 273 to 277 gussetts and nailing details.

Assembling and erecting the trusses

When considering practical details of construction there are one or two points worth noting.

1. It may be an advantage to exceed the minimum permitted spacing of nails and increase the sizes of gussets in order to give a greater rigidity and lateral strength when hoisting.

2. It is important to set the grain of the outer plies of the gussets roughly in the direction of the maximum stresslines so that the outer and alternate intermediate plies (of which overall there is one more) can take the compression and shear on the end grain.

3. It will be noticed that all nails in the rafters (Figs. 273 to 275) are kept 50 mm below the top edge so as not to foul the roofing batten nails.

4. Although the joint at B (Fig. 274) is under compression only, and theoretically only needs to be kept in position, larger gussets nailed on as shown add considerably to the stiffness. This is desirable for safe manhandling.

Trussed rafters should not be cut away at the feet (birdsmouthed). The rafters should be stored flat on site, clear of the ground and protected against the rain. The moisture content should not be allowed to exceed 22 percent.

When erecting, the trusses should be handled carefully to avoid disturbing the joints. Their positions should be set out accurately and the trusses fixed to the wall plates with 2 100 mm by No. 7 gauge skew nails to each joint. They should be carefully plumbed and braced and set at the apex with a temporary batten marked to the correct gauge. Unless the end rafters come tight to substantial gable end walls the roof should be braced by diagonals nailed to the undersides of the rafters extending over at least four spacings from wall plate to apex.

Gang-nailed trussed rafters have to be assembled in jigs with special cramping equipment and are made as standard proprietary units to various spans and pitches, prototypes of which are submitted to the appropriate authority (in this country, generally the Forest Products Research Laboratory) for tests. The design is therefore not the concern of the site carpenter. Figs. 278 to 280 give details of a typical truss of 25 deg. pitch suited to a span of 7.5 m.

Where possible, trimming for chimneys should be arranged to come within the rafter spacing; but where this is not possible, it is generally permitted to trim back one truss as shown in Figs. 281 and 282. If more than one truss has to be trimmed, the supporting trusses on either side may have to be strengthened.

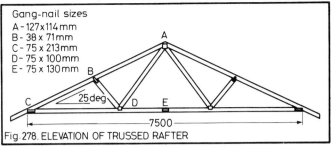

Gang-nail sizes
A - 127 x 114 mm
B - 38 x 71 mm
C - 75 x 213 mm
D - 75 x 100 mm
E - 75 x 130 mm

25 deg

7500

Fig. 278. ELEVATION OF TRUSSED RAFTER

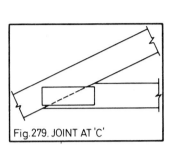

Fig. 279. JOINT AT 'C'

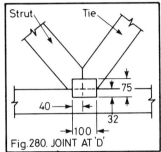

Strut   Tie

75
40   32
100

Fig. 280. JOINT AT 'D'

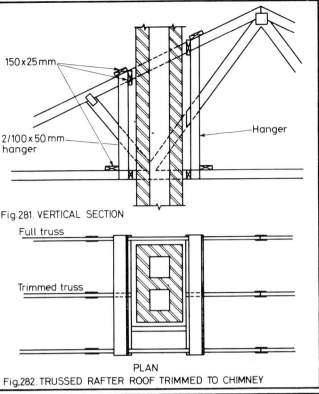

150 x 25 mm

Hanger

2/100 x 50 mm hanger

Fig. 281. VERTICAL SECTION

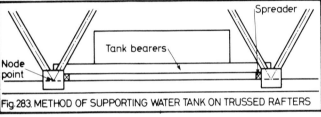

Full truss

Trimmed truss

PLAN
Fig. 282. TRUSSED RAFTER ROOF TRIMMED TO CHIMNEY

Spreader

Tank bearers

Node point

Fig. 283. METHOD OF SUPPORTING WATER TANK ON TRUSSED RAFTERS

Water tanks are better supported on load-bearing partitions rather than by the trussed rafters; but if this is unavoidable, the load should be distributed over a number of trusses by means of spreaders taken close to the node points (Fig. 283). It may be necessary to design trusses to carry the extra load or the spacing of the trusses may be reduced by introducing an extra one within the length of the spreader supporting the tank.

The popular demand for an extra room in the roof has created an interest in the possibility of designing trussed rafters with a clear central area to contain an attic with or without the support of a load-bearing partition. Tests have been carried out by the Forests Products Research Laboratory and these are now being manufactured by several firms, using Gang-nails or plywood gussets for the joints. Fig. 284 is the elevation and Figs 285 and 286 show typical details, although the sizes are only approximate.

It will be appreciated that the lower part forms the trussed unit, the upper part acting virtually as a separate couple-close roof.

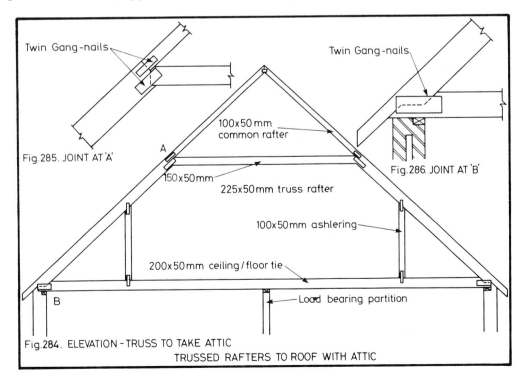

Twin Gang-nails

Twin Gang-nails

Fig. 285. JOINT AT 'A'

100 x 50 mm common rafter

A
150 x 50 mm
225 x 50 mm truss rafter

Fig. 286. JOINT AT 'B'

100 x 50 mm ashlering

200 x 50 mm ceiling / floor tie

B   Load bearing partition

Fig. 284. ELEVATION - TRUSS TO TAKE ATTIC
TRUSSED RAFTERS TO ROOF WITH ATTIC

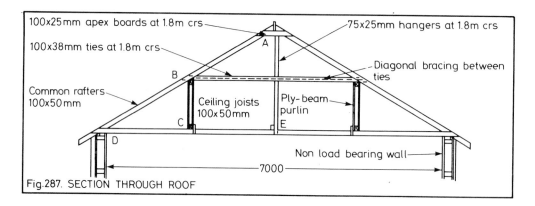

Fig. 287. SECTION THROUGH ROOF

The truss is incompletely triangulated, but distortion can only be produced by unequal forces, such as the wind acting on one slope, the effect of which would be to put one row of verticals forming the ashlaring into compression and the others into tension. These stresses must be resisted mainly by the stiffness of the joints and of the lower tie which is also subject to bending from normal floor and ceiling loads. A central load-bearing partition will considerably stiffen the truss and enable lighter members to be used.

Cross-wall construction is essentially a method of carrying all the weight of the roof and of the floors on transverse walls, mainly by means of trussed or boxed purlins, the top chord providing the normal bearing for the rafters and the bottom chord acting as a ceiling binder. As no weight is carried by external walls, these may be non-load-bearing and can if required by fully glazed.

Fig. 287 shows a section through the roof; although again, sizes are only approximate. It will be appreciated that the lower part of each rafter with the ceiling joist acts as a cantilever from the trussed purlin and that the rafter is in tension and the ceiling joist in compression (opposing loads balancing each other), as well as each being subject to the usual bending moments.

Unequal loading, such as wind loads, can cause distortion and produce lateral stresses in the purlins. These can be designed to resist this by making the upper and lower chords wide and flat or they can be more positively countered by transverse ties at, say, every 1.8 m with diagonals between.

The ceiling joists must be securely nailed to the feet of the rafters which must not be cut away (birdsmouthed). The purlins must be well supported and securely fixed to the cross-walls. If horns are left on the chords, these may be built into the walls; alternatively the purlins may be supported on steel channels as in Figs. 288 and 289.

Apex boards and central hangers should coincide with the horizontal ties at 1.8 m intervals. The rafters should not be notched over the purlins, but given a seating on continuous triangular section bearers, as shown in Fig. 290.

Figs. 291 to 294 show general details of construction and are self-explanatory.

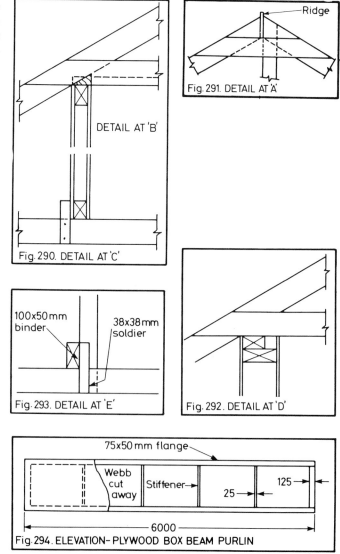

Fig. 290. DETAIL AT 'C'

Fig. 291. DETAIL AT 'A'

DETAIL AT 'B'

Fig. 293. DETAIL AT 'E'

Fig. 292. DETAIL AT 'D'

SUPPORT TO PURLIN FROM CROSS WALL
Fig. 288. END SECTION   Fig. 289. SIDE VIEW

Fig. 294. ELEVATION – PLYWOOD BOX BEAM PURLIN

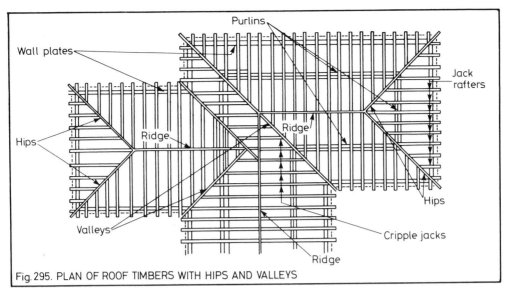

Fig. 295. PLAN OF ROOF TIMBERS WITH HIPS AND VALLEYS

Structural problems of hipped roofs

The construction of hipped roofs, which are not as popular as they were, involves some structural problems: the solution of which is largely based on experience rather than structural analysis. These problems are due to the diagonal thrust of the hips (and valleys) and also that of the hipped end jack rafters which cannot be tied back by the general line of ceiling joists, as can the rafters at the sides.

Fig. 295 is a plan (not to scale) of hipped roofs meeting at right angles to form valleys at the intersections. Where these roofs are of equal span (and pitch), ridges will meet at the same level with full length valleys; but when one span is reduced, its ridge will be lower. This is shown on the left of the drawing, one valley being taken past to the main ridge and the other butted against it to give the lower ridge level.

Jack rafters are birdsmouthed over the wall plates and splay cut against the hips. Cripple jacks are cut to the

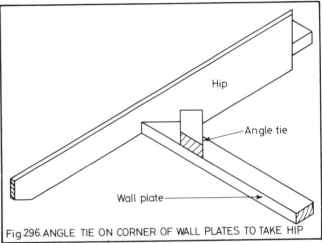

Fig. 296. ANGLE TIE ON CORNER OF WALL PLATES TO TAKE HIP

required bevel against the ridge and splay cut against the valleys. It is a moot point whether the hips carry the jack

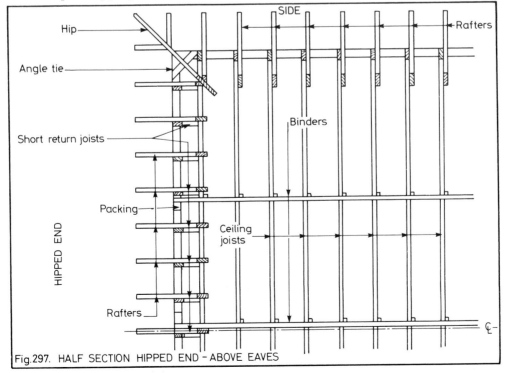

Fig. 297. HALF SECTION HIPPED END – ABOVE EAVES

71

rafter or the jack rafters carry the hips, but the hips do carry a load from the ends of the purlins, so there is likely to be a diagonal thrust at their feet. This used to be resolved by an elaborate formed beam-cum-tie known as a dragon beam, but this has been replaced for normal domestic work by an angle tie shown pictorially in Fig. 296.

The other problem which is not always dealt with as efficiently as it might comes from the outward thrust of the jack rafters against the wall plate at the hipped end. A part plan (Fig. 297) shows how this may be overcome by taking the last ceiling joist one space away from the end plate and nailing short joists at right angles to these and to the plate and jack rafters.

The ceiling binders are then extended right to the hipped end and used by whatever means presents itself to retain the end plate. Valleys take the feet or bottom ends of cripple jacks and it may be presumed take a greater load than the hips.

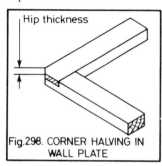

Fig.298. CORNER HALVING IN WALL PLATE

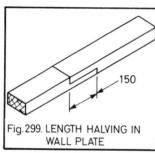

Fig.299. LENGTH HALVING IN WALL PLATE

There is no problem with the thrust at the feet of the valleys as this is taken by the combined resistance of the wall plates. They do, however, need strutting from a suitable point at mid-span to a convenient bearing. Figs. 298 and 299 show normal methods of joining wall plates.

PRACTICAL GEOMETRY APPLIED TO SIMPLE ROOFING

Most textbooks on carpentry and joinery or building, cover the geometry of roofing often quite extensively, as well as the construction and finishing of the roof itself. But few, if any, relate the one to the other in a practical way. Consequently, the budding craftsman comes on the site armed with a line drawing of basic lengths and bevels and does not know where to start. A preliminary check-up may tell him that the building is slightly out of square, while dimensions in general do not quite agree with the architect's specifications. This leads to more indecision.

This section aims to dispel this confusion by dealing with those points which hitherto appear to have been neglected in general written instructions.

In my opinion it is best to study the subject of roofing geometry (or for that matter any other building geometry) in three stages, as follows:

1. To think of the roof only as a series of lines and planes without measurable thicknesses; and, by means of accurate geometrical drawings aided perhaps by calculations, to find the true lengths of the lines and the magnitude of the angles at which the planes intersect.
2. To relate these bevels and dimensions to the solid timber units, and make adjustments in lengths to

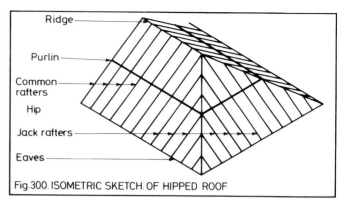

Fig.300. ISOMETRIC SKETCH OF HIPPED ROOF

accommodate the relevant thicknesses of intersecting members.
3. To make site checks and modify actual cut lines to take in site discrepancies. It may be presumed that the errors in actual lengths will not normally be large enough to affect appreciably the original setting-out bevels.

The various members in a hipped roof are shown pictorially in single line in Fig. 300 and in orthographic projection in Figs. 301 to 303. Though the diagrams do not show it entirely, it is as well at this initial stage to appreciate:

1. That the ridge, purlins, and wall plates are all horizontal and parallel to each other.
2. That the sides of the rafters, hips, ridge and wall plates are all in vertical planes.
3. That the face edges of all the rafters and the faces of the purlins are parallel to the roof slopes or planes.
4. That the sides of the purlins are square to the roof slopes.
5. That if the jack rafters are equally spaced they must diminish an equal amount one after the other.

Figs. 304 to 306 and 315 together show all the bevels necessary to cutting the timbers for a square plan hipped roof with equal pitches. They could all be contained in one diagram as in Fig. 318, but have been separated initially to avoid the complexity created by the crossing of too many lines.

Fig. 304 starts with a plan and vertical cross section through the roof. Bevels are indicated related to a schedule, also shown. A and B are common rafter plumb and seat cuts, the rafter lengths also being shown. The pitch of the

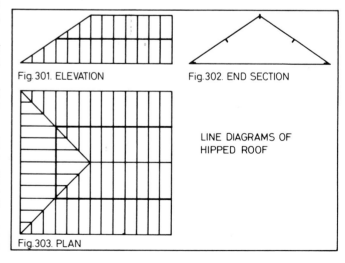

Fig. 301. ELEVATION

Fig.302. END SECTION

LINE DIAGRAMS OF HIPPED ROOF

Fig.303. PLAN

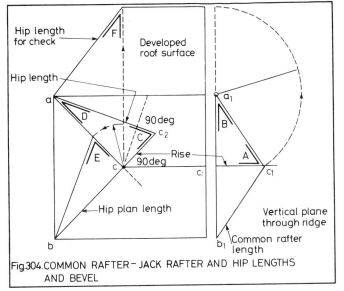

Fig.304. COMMON RAFTER – JACK RAFTER AND HIP LENGTHS AND BEVEL

roof which governs these values, may be given in one of two ways, either as an angle to the horizontal in degrees, or as the ratio of the rise to the span. It should be noted that the pitch is not given in relation to the run or plan length of the single rafter, but double that. Thus a half pitch roof will have a rise equal to the run, so the pitch will be 45 deg.

If the pitch is given in degrees, and the rise is not obvious as above, it is more accurate on a large scale to calculate the rise from the trigonometrical tables. Thus for a roof of, say, 10 m span at 40 deg. pitch, run = 5 m and rise = tan. 40 deg. × 5 m = 0.8391 × 5 = 4.176 m.

It is possible to use trigonometrical tables in roofing with only a very basic knowledge of the subject and it may help the reader if he thinks of the trigonometrical ratios in roofing terms.

Let the angle X = the pitch of the roof in degrees. Then, taking names for lengths

$$\text{Sine X} = \frac{\text{rise}}{\text{rafter.}} \quad \text{Cosine X} = \frac{\text{run.}}{\text{rafter.}} \quad \text{Tangent X} = \frac{\text{rise}}{\text{run}}$$

$$\text{Cosecant X} = \frac{\text{rafter}}{\text{rise}} \quad \text{Secant X} = \frac{\text{rafter}}{\text{run}} \quad \text{Cotangent X} = \frac{\text{run}}{\text{rise}}.$$

From this it will follow that:
a. Given the run, rise = Tan. X run. Rafter length = Cos. X run.
b. Given the rise, run = Cotan X rise. Rafter length = Cosec X rise.

c. Given the rafter length, rise = Sin. X length, run = Cos X length

In Fig. 304, 'a-c' is the plan length of the hip which stands the distance of the rise above its plan at the centre. Thus the right angled triangle 'a-c-c₂' contains the hip bevels 'C' and 'D' and the hip length.

A glance at Fig. 300 should make it clear that if the roof side were hinged at the eaves to lie flat it would take the shape shown in Fig. 304 containing jack rafter edge cut.

As the jack rafters' edges lie in the roof plane and intersect the vertical plane of the hip, the two planes must intersect vertically so that the jack rafter plumb cuts will be as for the common rafter. The top edge of the hip left in the square state does not lie in the roof planes, and in this condition will have to support the tiling battens on its sharp corners. As there is little weight transmitted by individual roofing battens, this is often regarded as satisfactory, although the purist will insist that the hip edge is double bevelled to lie in the corresponding roof planes. If the roof planes are of unequal pitches, this bevelling becomes a necessity as the arrises would lie at different levels with regard to the relative roof planes.

In order to understand the geometry of this bevel, known as the backing bevel, it is necessary to appreciate that it is formed at the intersection of four planes. These are the two roof planes, the vertical plane containing the centre of the hip and an inclined plane on which the bevel is actually set which is square to the hip plane on plan, giving a horizontal trace b-c and tilted so that it is square to the hip line in the vertical plane as shown by the dotted line at 90 deg. to the hip.

In the drawing Fig. 304; both hip plane 'a-c-c₂' which is hinged about line 'a-c' and the plane containing the backing bevel 'E' which is hinged about line 'b-c' are shown flat on the deck. They should be visualised as lifted into their original positions.

If this drawing is reconstructed on stout paper and cut along the profiles, the two planes can be folded up and the developed roof surface folded over to meet them.

Two bevels are needed to cut the purlins and are shown in Fig. 305. If the purlin stands below the hip, it has to be cut to fit under it, using the lip cut as shown in Fig. 306.

Referring to Fig. 305, the section on the right shows the edges of the roof planes which are presumed to pass under the rafters and contain the faces of the purlins. It also contains the sections of the purlins, the back and bottom surface of one (not necessary to the bevel development) being omitted for the sake of clarity. If these purlins were

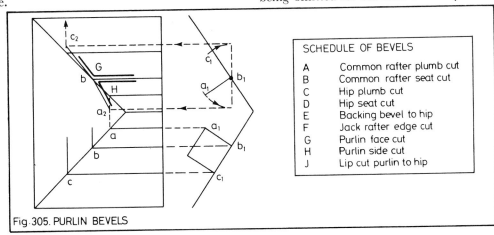

Fig.305. PURLIN BEVELS

SCHEDULE OF BEVELS	
A	Common rafter plumb cut
B	Common rafter seat cut
C	Hip plumb cut
D	Hip seat cut
E	Backing bevel to hip
F	Jack rafter edge cut
G	Purlin face cut
H	Purlin side cut
J	Lip cut purlin to hip

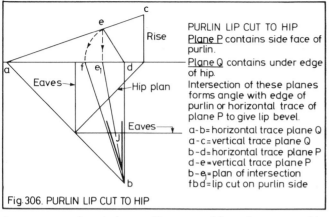

PURLIN LIP CUT TO HIP

Plane P contains side face of purlin.

Plane Q contains under edge of hip.

Intersection of these planes forms angle with edge of purlin or horizontal trace of plane P to give lip bevel.

a-b = horizontal trace plane Q
a-c = vertical trace plane Q
b-d = horizontal trace plane P
d-e = vertical trace plane P
b-e₁ = plan of intersection
fbd = lip cut on purlin side

Fig.306. PURLIN LIP CUT TO HIP

drawn to scale, their smallness would make reasonable accuracy impossible; they have, therefore, been greatly enlarged.

At the lower part of the drawing, the plan of the intersection of the arrises with the hip is shown in points 'a,' 'b' and 'c' and the corresponding points in the section in 'a_1,' 'b_1' and 'c_1'. This is also shown but not referenced in the top part of the section. If the purlin faces are imagined as being hinged about the arrises 'b' or 'b_1' into the horizontal plane, then this movement will be shown in the section by arrowed arcs which limit the extent of the movement, and on plan in an outward direction terminating in 'a_2' and 'c_2'.

To obtain the lip cut Fig. 306, it is necessary to find the intersection of the plane containing the underside of the lip (vertical trace 'a-c', horizontal trace 'a-b') and the side face of the purlin horizontal trace 'd-b' and the vertical trace 'e-d'. The intersection line must be from point 'e' ('e_1' on plan) and 'b'.

To get the true bevel this must be hinged at 'b-d' and folded down to 'f' to give the triangle 'b-d-f' containing hip cut 'j', applied off the bottom edge of the purlin.

The only other bevels required are the mitres on the hip which, with the plumb cut give the intersection joints at the ridge. These have been omitted at this stage as they are not strictly necessary and the marking out will be shown without their direct application, but they will be given later.

Relating bevels and dimensions to timber units

It now becomes necessary to relate the lengths and bevels directly to the roofing timbers and make the necessary allowances for the member thicknesses.

First of all we have to be quite clear as to where the basic lengths shown on the scale drawing, lie within the roof

structure, individual craftsmen have different ideas about this but I have always taken them as shown in Fig. 307, the basic lines being the heavy ones.

It will be seen that the span is taken as width over wall plates, the rafter length is measured along its top edge terminated by the span as shown. The rise corresponds with this within the closed triangle.

The hip line will go from a point distance 'D' vertically above the uncut corner of the wall plate in a vertical plane at 45 deg. to the wall plates to intersect the centre line of the ridge. The theoretical position of the rafter will be the rafter run (half span) back from the end corner of the plate.

In marking cut roofing timbers it is generally best to start by marking the basic length from the drawing on the timber and work from this to give adjustments for timber thicknesses and modifications for site discrepancies.

In all setting out for roofs (or for that matter any splayed or bevelled in situ work), the following should be emphasised and never forgotten:

If any measurement is made square to a plumb cut it must necessarily be a plan measurement.

This means that any length shown on the plan can be transferred direct to the vertical face of, say, the side of a rafter or hip by measuring square to the plumb cut. The squareness of the measurement can quite safely be estimated with the eye with normal care.

To mark out the common rafters, a pattern rafter should first of all be very carefully marked out, cut, and very clearly identified and all the others marked from it. The continuous marking of 'one from the other' will lead to an accumulation of errors. It may be an advantage in the interests of accuracy and convenience to mark out and cut a thin, straight planed board of the requisite width and use this for a pattern. Fig. 308 shows the set out.

The basic length, indicated by the thick line, is marked on the timber and the plumb bevels pencilled in at both ends. At the top end, half the thickness of the ridge is set back measured square to the plumb line to give the actual plumb cut (adjustments for discrepancies where necessary will be discussed later). At the bottom end the depth of the birdsmouth is marked as shown, the usual maximum depth of the birdsmouth being as in Fig. 307. The distance 'D'

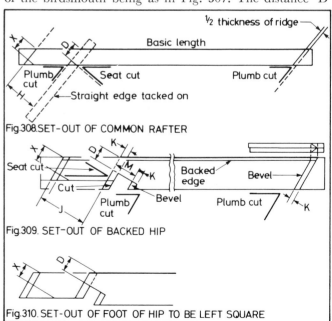

Fig.308. SET-OUT OF COMMON RAFTER

Fig.309. SET-OUT OF BACKED HIP

Fig.310. SET-OUT OF FOOT OF HIP TO BE LEFT SQUARE

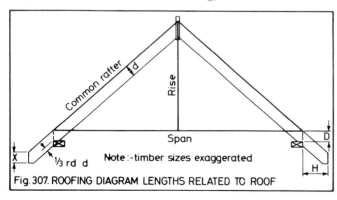

Note:- timber sizes exaggerated

Fig.307. ROOFING DIAGRAM LENGTHS RELATED TO ROOF

74

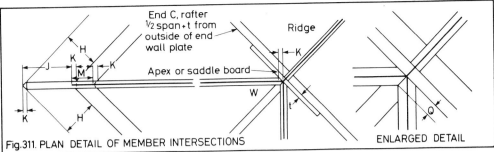

Fig.311. PLAN DETAIL OF MEMBER INTERSECTIONS   ENLARGED DETAIL

now becomes a constant and represents the vertical height of the roof plane to face edges of rafters over the outside edges of wall plates at any point.

The feet of the rafters, giving eaves overhang, may be cut after fixing; but this is much more easily done on the bench. If a straight edge is lightly tacked on against the plumb cut, the required plan distance can be measured square from this for the eaves fascia line. Note that the distance 'X', giving the soffit level, also becomes a constant for rafters, hips and valleys.

To mark out the hips, again start with the basic length on the top edge with the plumb bevel at each end as in Fig. 309. Adjustments at the apex level will be governed by roof construction and arrangement of members around the intersection between rafters, hips and ridge.

Unless one is familiar with the method used, it is better to set out the plan of the intersection full size as in the right of Fig. 311. The basic intersecting centre lines (shown heavy) should be drawn first and then, as it were, clothed in timber, this will show adjustments which have to be made off the plumb lines.

The detail here will also govern the position of the first pair of common rafters to be marked on the side wall plates. In the example given it should be appreciated each will be set back half the centre line (basic) span + thickness of saddle board from the outside of the end plates.

The edge bevel, which with the plumb cut forms the hip mitre, shows as a mitre on plan and therefore may be marked out by a parallel half the hip thickness from the original plumb line, continuing to the top squareing over as shown to give the mitres either on the square or backed edge.

The cut over the wall plate at the foot of the hip is obtained by measurements 'M' and 'K' taken square off the plumb line, the inside cut being mitred and the outside squared to fit over the stubbed off corner of the plate as shown to a larger scale in Fig. 312.

The depth of the horizontal seating on to the wall plate as given by the seat cut is again related to the constant

'D' and depends upon whether the hip is to be backed or unbacked. If it is to be backed the measurement 'D' is down the plumb line from the end of the original basic line and the arrises of the hip will stand above the rafter line until bevelled off.

If the hip is to be left square, the measurement is taken down the cut line as in Fig. 310. It will be seen that this drops the hip lower so that the corner of the hip edge lies in the required roof plane.

The eaves overhang is marked in the same way as for the common rafter, but the overhang at the longest point (on plan) is the diagonal of the eaves overhang squared, or common rafter eaves overhang × 1.414 so that if the rafter overhang = 300 mm, hip overhang = 424.2 mm (estimating the 0.2 mm gives a high degree of accuracy).

It will be noticed that half the hip thickness represented by the constant 'K', is used for all the hip cuts showing a mitre on plan. If, however, the reader prefers to use a set bevel for this then the bevel to be placed on the top edge of the backed hip is the jack rafter edge bevel ('F' in Fig. 304); but if the top edge is to be left square, a special bevel is needed.

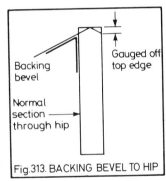

Fig.313. BACKING BEVEL TO HIP

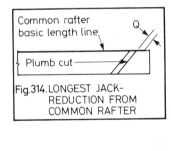

Fig.314. LONGEST JACK-REDUCTION FROM COMMON RAFTER

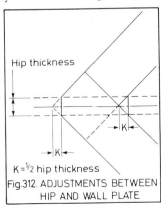

K = ½ hip thickness
Fig.312. ADJUSTMENTS BETWEEN HIP AND WALL PLATE

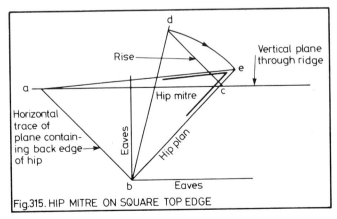

Fig.315. HIP MITRE ON SQUARE TOP EDGE

This is given in Fig. 315 by the intersection between the plane containing the back edge of the hip, its horizontal trace being square to the hip plan, and the vertical planes through the centre of the ridge and the centre of the hip. The three planes intersect at point 'c' in plan corresponding with point 'd' on the vertical section through the hip shown laid flat.

If the plan 'a-b-c' containing the required bevel is folded down about the horizontal trace 'a-b', 'a-c' will extend the full hip length to 'e' to give the required bevel. The backing bevels to the hip are planed off the top edge corners to the centre, the depth of the bevel being gauged off the top edge. The distance to be gauged each side is given in Fig. 313.

In marking out the jack rafters, the longest jack is set out first. This is shortened from the common rafter by the amount 'Q' (enlarged detail Fig. 311) measured from the plan and set off square to the plumb cut. All successive jacks shortened progressively by a regular amount, governed by rafter spacing.

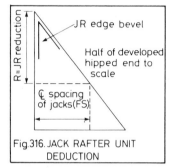

Fig.316. JACK RAFTER UNIT DEDUCTION

Fig. 316 shows how this constant is obtained. A line is drawn on the developed roof plane parallel to the longest jack or first common rafter and a distance from it equal to the rafter centre line spacing. A horizontal drawn back then gives the reduction 'R'. The dotted lines show this.

The question of accuracy is an important one in roofing where measurements are taken from a scale drawing. The scale, and consequently the overall size of the geometrical set-out, is governed by the size of the plane surface conveniently available for the drawings. It is an advantage, therefore, to be able to make these as compact as possible.

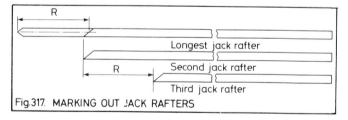

Fig.317. MARKING OUT JACK RAFTERS

Fig. 318 shows a method of doing this containing all the lengths and bevels within the smallest possible rectangle in relation to the scale. The reader may now like to compare it with the previous drawings, Figs. 304, 305 and 315 (the angles are identified with the schedule but the lip cut has not been included).

The drawing is not so simple to understand and does not follow all the rules of orthographic projection. The principle on which it is constructed is that certain lengths represent certain values irrespective as to their plan positions and can be used in any convenient order. Thus 'a-b', 'b-c', 'c-d', or 'd-a', can be the rafter run and 'a-o-c' or 'b-o-d' can be the hip run. This method can only be used for equal pitched roofs on rectangular plans. All the work is confined to one half of the hipped end.

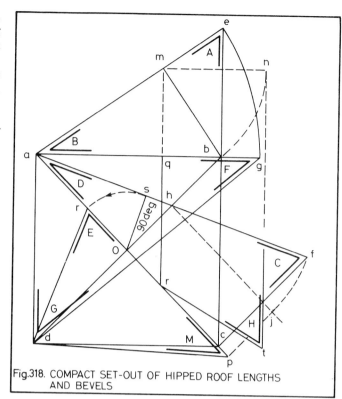

Fig.318. COMPACT SET-OUT OF HIPPED ROOF LENGTHS AND BEVELS

The rise 'b-c' plumbed off the run 'a-b' gives common rafter bevels 'A' and 'B' and the length. This length is hinged down to give the developed roof surface with the jack rafter bevel 'F'. As this plane at the eaves is parallel to the purlin edges, the bottom bevel 'G' becomes the purlin face bevel. The rise 'c-f' squared off the hip plan 'a-c' gives the hip length 'a-f' and plumb and seat cuts 'C' and 'D'.

Half the length of the hip 'h-f' (plan 'o-c') hinges on the trace 'o-d' and folds down to give the hip mitre. (Compare relative lengths with Fig. 315.) The plane containing the purlin side, placed for convenience at 'm-b' in elevation and the 'q-r-b-c' in plan, swings up to the horizontal in 'm' in the vertical plane and out to 't' in plan to give the purlin side cut in 'H'. The backing bevel is the same as 'E' in Fig. 305, but taken in a different position.

It may now be worthwhile to calculate what degree of error is likely to occur in marking out from the scale drawings. Take an average span of, say, 8000 mm with a pitch of 35 deg., then from Fig. 318 the set-out is going to cover about 8000 by 6000 mm to scale. Assuming a sheet of plywood 1200 by 2400 mm was available for the drawings, the largest possible scale would be

$$\frac{1200}{6000} = \frac{1}{5}.$$

As the initial dimensions for set-out are only rafter run and rise, any other scale with a numerator of 1 could conveniently be used, say,

$$\frac{1}{6}, \frac{1}{7} \text{ or } \frac{1}{8}.$$

The measured lengths would only have to be multiplied by the denominator to get the end result.

Assuming the scale is 1 : 8, then an error of 1 mm in the plan would give one of 8 mm on rafter length. Assume the rafter is 4500 mm long, then the measured length may be

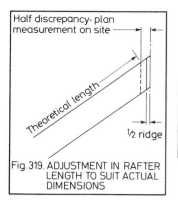

Fig. 319. ADJUSTMENT IN RAFTER LENGTH TO SUIT ACTUAL DIMENSIONS

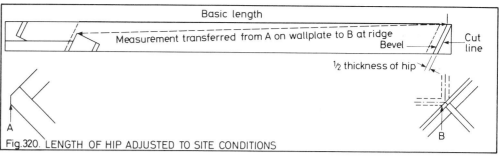

Fig.320. LENGTH OF HIP ADJUSTED TO SITE CONDITIONS

4508 mm. This will result in a slight increase in pitch, but will this be sufficient to affect the fit of the joint?

Using the trig. tables, the change in pitch may be calculated as follows:

Run of rafter at 4500 length = cos. 35 deg. × 4500 = 0.8192 × 4500 = 3686.4 mm.

Cos. new pitch:

$$= \frac{\text{run of rafter}}{\text{new length}} = \frac{3686.4}{4508} = 0.81775 = \cos. \text{ angle } 35.15$$

deg.

The error is only about 1/7 degree which will not appreciably affect the fit of the cut. The rafters must, of course, all be the same length. There may still be discrepancies on site; but if adjustments are made for these, the fit of the joints will not be affected at all.

Fig. 319 shows how adjustment may be made. Assuming span checked over wall plates is plus 50 mm. Half this measured square off the plumb line from the basic length gives the modified centre line length. The cut line is then measured half the ridge thickness back from this. The matter is not so simple, however, with the hip rafter as the building may also be out of square meaning that the hips will vary in length. The only answer is to check the actual hip length on site from wall plate to ridge. The hip should first of all be set out complete, as if all dimensions fully conformed and, then measurements taken from the shorn corner of the wall plate (Fig. 320) 'A' to the actual point 'B' at the apex where the hip has to fit, and transferred to the corresponding points on the hip. The hip mitre has then to be added on.

By and large, the techniques described here only relate to normal hip roof construction with equal pitches to all slopes and the roof plan rectangular. Irregular roof plans pitches require different treatment.

THE CARPENTER'S STEEL SQUARE

A great deal was said and written about the carpenters' steel square in the days of imperial measure and many magical qualities attributed to it. In my opinion, however, it is primarily an instrument for marking a standard hipped roof, square on plan, with equal pitches, without first having to make a drawing or for any other carpentry work involving similar simple angular construction. For anything more complex than this some drawing board work is essential and rather than mix up the two systems it is better to stick to the drawing board.

The instrument is also a useful tool to have in a bag as a square in its own right.

The original steel square in imperial measure contained various tables for lengths taken on the square to give bevels and length per foot run to standard pitches.

There is one metric square produced by Smallwood (there may be others) now on the market which besides having the metric equivalent of the old tables also gives scales for setting up various angles in degrees. In my opinion one fault of the old square was that it was rather clumsy to use being 3 to 5 mm thick and necessarily square edged it was prone to errors in reading due to parallax. If subject to hazards commonly met on the building site liable to lose its legibility. For anyone buying one, together with the usual instruction booklet it will undoubtedly prove itself a useful tool and may enable the user to carry out certain operations without necessarily understanding them.

It is however, in my opinion better to have a sound knowledge of the principles behind the use of the square then use the tables afterwards for convenience.

The following, therefore is confined to explaining these principles, treating the square purely as a graduated right-angled triangle, applying them only in relation to a hipped roof, rectangular on plan and with equal pitches.

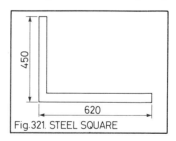

Fig. 321. STEEL SQUARE

Fig. 321 is the outline of a steel square, the wide part being termed the blade and the narrow part the tongue. It is presumed to be graduated in mm.

The simplest way to understand the square is probably to think of it as being applied to a model roof of 1 : 10 scale. The bevels will, of course, still be the same as for the full-sized roof, but the lengths will have to be multiplied by 10. Calculations applied to the square have been greatly simplified with the introduction of SI units.

For example, having obtained the run of the rafter or half span of the roof this is divided by 10 to give the dimension for the square. Assuming the run is 3947 mm, then the value to take on the square is 394.7 mm, the 0.7 mm being estimated with the eye. The rise may be given either in mm, using the same procedure as before, as a ratio (say 1 : 3) in which case the span (not the run) is divided by 3 and then by 10; or in degrees, in which case the actual rise

ROOFING

may be obtained from a scale on the square or by using trigonometrical tables.

If the angle is say 35 deg. then the rise will be tangent 35 deg. X run, or taking the above example $0.7002 \times 3947 = 2764$ mm or 276.4 taken on the square. The rise and run are the only dimensions needed to obtain the basic lengths and bevels. All the other requirements are evaluated in the process of setting out.

There is, in my opinion, a need to maintain the maximum accuracy. To this end, all the initial setting out of lengths and bevels should be done on a smooth board with a shot edge. It is better if the lengths and bevels are all marked on the board and clearly identified as they are obtained. It is a mistake to mark scale dimensions on the sawn timber and even worse to step them, say 10 times down the timber to arrive at the ultimate length.

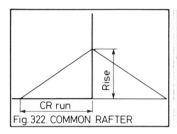
Fig. 322. COMMON RAFTER

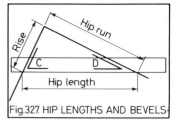
Fig. 325. OBTAINING HIP RUN

The basic lengths are initially marked on the timber, adjustments being made for thickness from the plumb line as previously described. Fig. 322 shows the square in outline pictured as placed in the 1 : 10 scale roof section. In Fig. 323, the square is turned over and laid on the board to give the plumb and seat cuts 'A' and 'B' together with the scale rafter basic length which must be multiplied by 10 for the full dimension.

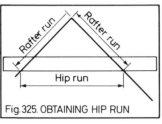
Fig. 323. COMMON RAFTER

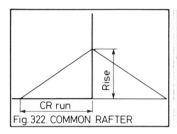
Fig. 324. HIP DETAIL

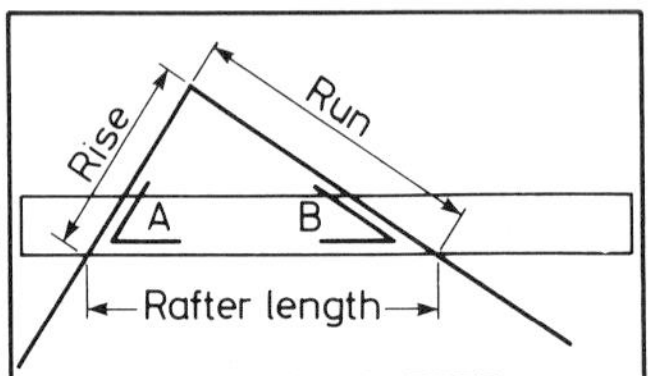

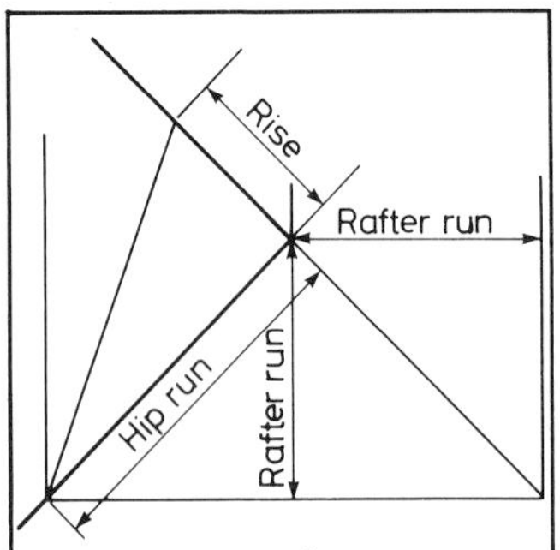

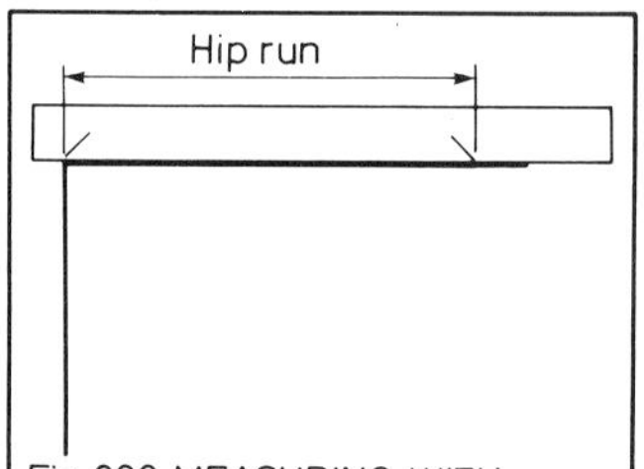

Fig. 326. MEASURING WITH SQUARE

In Fig. 324, the square is shown with the blade placed on the plan of the hip, the rise then taken on the tongue produces a triangle; the hypotenuse of which gives the scale hip length and the acute angles the plumb and seat cut. The hip run is the diagonal of the rafter run squared. This must be obtained first as shown in Fig. 325 and marked on the shot edge of the board from which it is taken as shown in Fig. 326.

Fig. 327 shows the square placed on the board and the plumb cut and seat cut 'C' and 'D' and the scale length

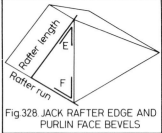
Fig. 327. HIP LENGTHS AND BEVELS

Fig. 328. JACK RAFTER EDGE AND PURLIN FACE BEVELS

marked off. Fig. 328 shows the square resting on the scaled down roof indicating why the rafter length and rafter run give the jack rafter edge cut as well as the hip mitre on the backed edge 'E' and the purlin face bevel 'F'.

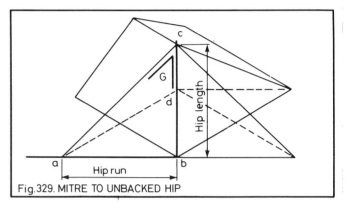
Fig. 329. MITRE TO UNBACKED HIP

Fig. 329 shows the square lying in a plane with the horizontal trace 'ab' cutting a vertical plane acd through the common rafter to give the hip mitre for unbacked edge 'G'.

The logic behind the purlin side bevel and the backing bevel for the hip is a little more complicated, but if the reader understands this he is likely to remember them.

Take the purlin side bevel, (Fig. 330), and consider it as being obtained on the drawing board first. The purlin is placed for convenience so that the plane containing its upper side continued will have its horizontal trace under

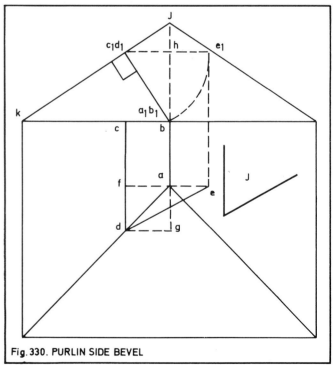
Fig. 330. PURLIN SIDE BEVEL

the ridge. We have to find the angle at which it cuts the plane of the vertical side of the hip.

The plan of this purlin plane will then be, 'abcd' and elevation 'a_i b_i c_i d_i'. If this plane is now hinged about 'cd' into the horizontal it will extend to 'e_i' in elevation and 'e' in plan, giving the angle 'cde' represented by the angle 'J' which is the angle we want.

The lengths therefore for marking on the square, if we knew them, would be 'fe' on the blade and 'fd' on the tongue. However, as 'fagd' is a square fd = fa = d_ih and $d_i b_i$ = fe so what could be used would be 'd_ih' on the tongue and '$d_i b_i$' on the blade or lengths proportional to them.

The triangles '$d_i h b_i$' and 'jb_ik' are similar so that 'd_ih : $d_i b_i$ :: jb_i:jk' which last is rise:rafter length. Therefore to obtain purlin side cut take rise on the tongue and rafter length on the blade.

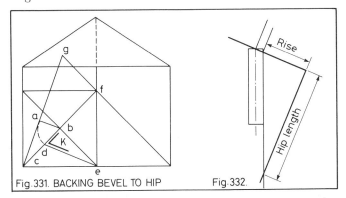

Fig.331. BACKING BEVEL TO HIP   Fig.332.

For the backing bevel to the hip Fig. 331 shows the geometrical method of obtaining this. For the square, this would need 'db' on the tongue and 'be' on the blade, but we do not have these, however 'db = ab' and 'eb = cb'. Both of these lie in the hip triangle and the triangle abc is proportional to the large triangle 'fgc' so that 'ab:bc :: fg:gc' = rise:hip length so that the backing bevel may be

obtained with rise on the tongue and hip length on the blade. Fig. 332 shows this in relation to the hip section.

Obtaining the lip cut — if the reader refers to the section dealing with simple roofs, he will appreciate that the lip cut is very nearly rise:2 × common rafter run. As the cut will only be half the hip thickness or 19 mm in length, these values are accurate enough. These can be more conveniently taken on the square as half rise on the tongue and rafter run on the blade.

When the square is placed on a roof member or on a board to obtain a bevel, two angles are, of course, always shown. The newcomer may be unsure of which one to take. The following facts (related only to the simple roof), if memorised, will help his confidence over this problem.

Common rafter and hip plumb and seat cuts — the plumb cut always comes against the rise.

Jack rafter edge cuts and hip mitre — these are always less than 45 deg.

Purlin side and face bevels — these are always more than 45 deg.

The backing bevel on the hip (ie. the half bevel not the full dihedral angle) is always more than 45 deg.

Lip cut of purlin under hip is always less than 45 deg.

ROOFS OF IRREGULAR PLAN AND UNEQUAL PITCHES

Although the majority of buildings are rectangular with equally pitched roofs, the need occurs occasionally to make maximum use of a sloping or limited site by designing houses of irregular plan and floor arrangement, therefore creating varied roof shapes and unequal pitches.

Starting with a simple example and dealing progressively with the more difficult, the first roof to be discussed is one of equal pitches but with an oblique hipped end. Fig. 333 is a line diagram showing the geometry and basic lengths of rafters and hips.

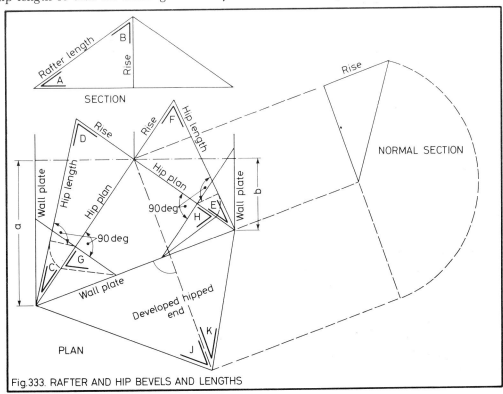

Fig.333. RAFTER AND HIP BEVELS AND LENGTHS

ROOFING

The first requirement in all pitched roofs, in order to avoid splayed edges to common rafters, is that the rafters shall be square to the wall plates on plan. Thus the outline plan and vertical section will give the common rafter lengths and bevels and the common rise, from which a normal section square to the end plate may be drawn and produced to give the intersection of the longest jack rafter and the ridge, and subsequently the hips.

As all pitches are equal, the plan of each hip must bisect the angle of the wall plates. From this it follows that, as the side wall plates are parallel, the sum of these angles must be 180 deg., the sum of the angles of the hips with the end plate half of this or 90 deg., and the intersection of the hips on plan (180 — 90) also 90 deg.

The main section gives the rafter length, the seat cut 'A' and plumb cut 'B' of all the common rafters and jack rafters. The edge cuts 'J' and 'K' of the jack rafters are obtained by developing the hipped end hinged about the end plate and projected from the rotated length from the normal section. Setting up the rise square to the hip plans gives hip lengths, seat cuts 'C' and 'E' and plumb cuts 'D' and 'F'.

The plane which contains the backing bevel has its horizontal trace square to the hip plan in each case. It cuts the vertical hip section square to the hip and may be swung down on to the plan to give the backing bevels 'G' and 'H' respectively.

For simplicity, the purlin bevels are shown separately in Fig. 334. The exaggerated purlins in the section are projected down to cut the hips, the mitre point being swung out to the limit dictated by the hinged-up faces in the section to give the purlin bevels 'M', 'N', 'P' and 'Q'.

To give the lip cut, where the purlin extends below the hip and is cut under it, the purlin may be presumed to be placed so that the plane of its top side extends down to from a horizontal trace directly under the ridge. The lip cut (applied off the bottom edge of the purlin to the side) is the angle at which this plane meets a plane containing the squared under edge of the hip (horizontal trace 'xa').

To obtain this intersection, a vertical plane 'xby' cuts through the hip intersections square to the ridge, the hip planes cutting this at 'x_ib_i' on the left and 'y_ib_i' on the right. The purlin planes can then be extended upwards to cut these lines at 'c_i' and 'd_i' ('c' and 'd' on plan). The purlin planes can then be hinged down about line 'ab' to give lip cuts.

The other (unidentified) bevels show how the same results may be more conveniently obtained by taking parallel planes square to the roof pitch wall plate level.

There are not many problems in setting out the roof timbers. It is better to draw full-sized plans of hips and ridge and also hip and wall plates as shown in Figs. 336 to 338, also a detail of the birdsmouth of common rafters with dimension 'd' in Fig. 339.

Figs. 340 and 341 show how dimensions may be taken from these and marked square off plumb cut lines placed on the timber from the basic geometrical drawings; they are squared over the edge, where necessary, to give points through which to mark mitre bevels.

It should be noted that the position of the first pair of rafters as marked on the side wall plates is governed by the distances 'a' and 'b', Fig. 333 measured from the hipped end. This must coincide with the centre line intersection of hips and ridge (Fig. 342). In whatever way it may be

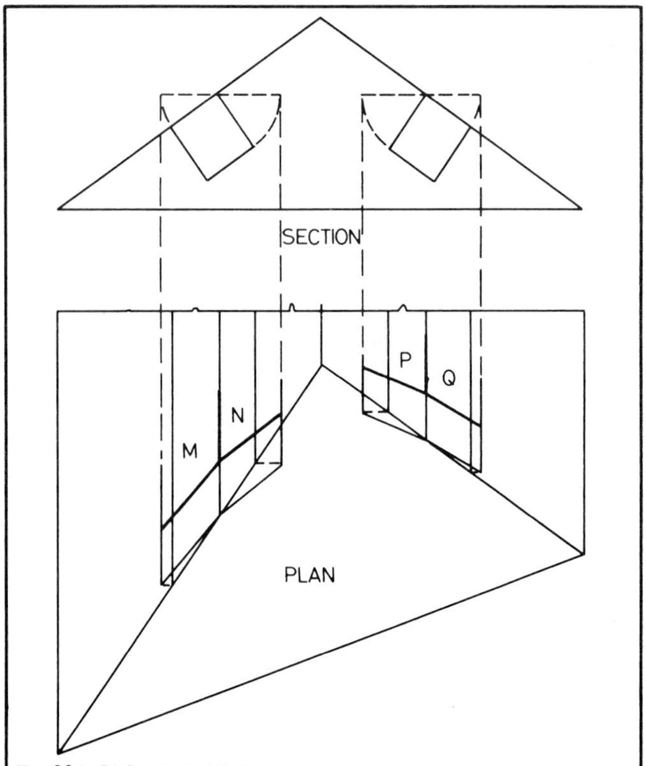

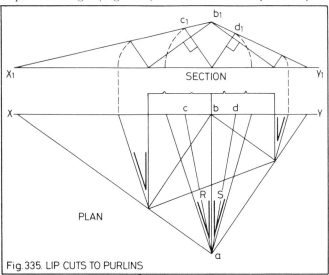

Fig. 335. LIP CUTS TO PURLINS

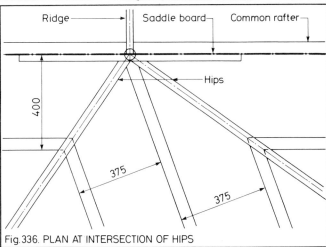

Fig. 336. PLAN AT INTERSECTION OF HIPS

Fig. 334. PURLIN BEVELS

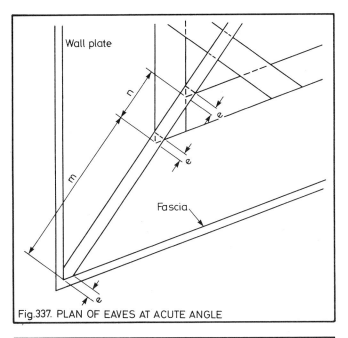

Fig.337. PLAN OF EAVES AT ACUTE ANGLE

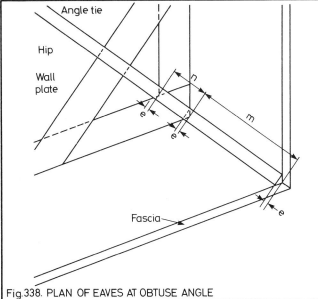

Fig.338. PLAN OF EAVES AT OBTUSE ANGLE

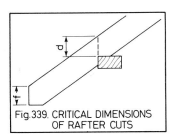

Fig.339. CRITICAL DIMENSIONS OF RAFTER CUTS

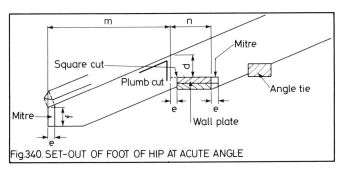

Fig.340. SET-OUT OF FOOT OF HIP AT ACUTE ANGLE

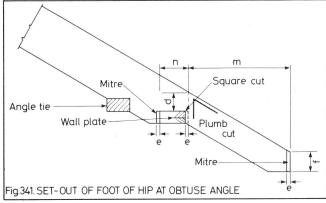

Fig.341. SET-OUT OF FOOT OF HIP AT OBTUSE ANGLE

decided to join these roof members, these centre lines must be drawn first on plan and the method of construction built around it, the first pair of rafters being adjusted accordingly.

In the example given, the centre line coincides with the face of the end rafter. Other details do not differ from standard construction and need no further explanation.

The set-out and construction of a roof on a rectangular plan, but with the hipped end at a steeper pitch, will now be considered. The basic geometry of this is shown in Fig. 343. The plan and main sections are drawn first.

The rise obtained from the section limits the length of the common rafter or longest jack drawn to a given pitch for the hipped end and enables the plan to be completed. It will be obvious that the hip does not bisect the angle of the wall plates. There will, of course, be two rafter lengths and two sets of rafter bevels 'A', 'B' and 'A$_i$', 'B$_i$'. The rise drawn square off the hip plan gives the hip length and the plumb and seat cut 'C' and 'D'.

The backing bevel or dihedral angle lies in a plane square to the hip in section and has its horizontal trace square to the hip plan and is swung down on to the horizontal plane about the horizontal trace to give the dihedral angle divided by the hip plan giving two angles, 'P' and 'M' applied to the hip as shown in the section.

The purlin bevels are obtained from the respective sections by swinging the faces about the edges into the horizontal plane and projecting down to limit the outer

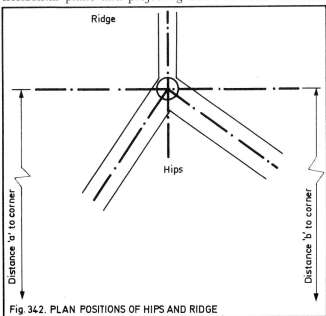

Fig. 342. PLAN POSITIONS OF HIPS AND RIDGE

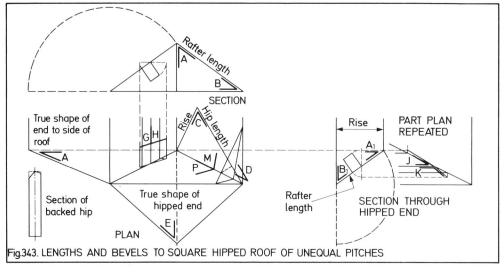

Fig.343. LENGTHS AND BEVELS TO SQUARE HIPPED ROOF OF UNEQUAL PITCHES

movement of the corresponding points in plan and give the bevels 'G', 'H', 'J' and 'K' required.

Part of the roof has been re-drawn to separate the developments and avoid confusion. The edge bevels of the jack rafters 'F' and 'E' are obtained from the developed roof side and hipped end respectively, the movement into the horizontal plane being shown in the sections by the arcs radiating about wall plate lines. The lip cut is the same as before for the side purlins; but for the end purlins, the 'xy' line is taken through the vertical plane of the ridge and looking towards the hipped end section.

Coming now to structural problems: first, as the hip does not mitre the plan of the wall plate lines, uniformly spaced jack rafters cannot meet on the hips. Thus, depending upon the differences in pitches, the horizontal thrust of some jack rafters will not be directly countered. However, as the dead load is primarily vertical and the hip will be tied at intervals of about 200 mm with tile battens, failure of the hip is unlikely.

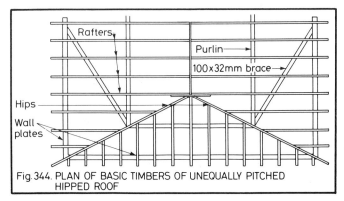

Fig. 344. PLAN OF BASIC TIMBERS OF UNEQUALLY PITCHED HIPPED ROOF

If the building is in an exposed position, it may be advisable to use 50 mm thick hips instead of the usual 38 mm. As an additional security against unequal stresses due to unequal pitches, it may be advisable to brace the centre of each hip. These details are shown in Fig. 344.

Another effect of unequal pitches, as far as design is concerned, is that, if standard construction is used, the overhang of the eaves will be less on the steeper pitched side of the roof; the amount depends upon the extent of the variation. Within certain limits, this can be adjusted by altering the positions of the wall plates in relation to the external walls, as shown in Fig. 345.

If this cannot be done within available limits, the wall plate will have to be taken higher at the steeply pitched end and connected to the lower one by built-in packings. Some ingenuity will be needed to provide a sound angle tie at the corner of the plates.

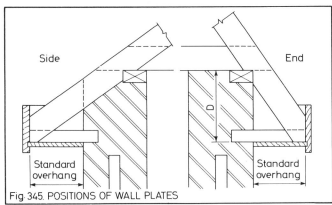

Fig. 345. POSITIONS OF WALL PLATES

Fig. 346 shows the set-out of the hip end and wall plate from the plan to be drawn full size, see also Fig. 344. It is obvious that the hip cannot fit squarely over the angle and it will have to rely on the angle tie for support. In marking out the members, the procedure already dealt with of marking plan measurements from the plumb lines should be used.

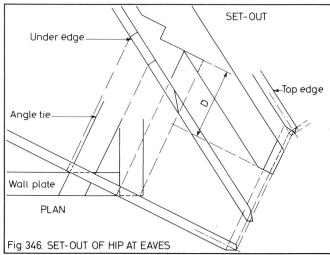

Fig. 346. SET-OUT OF HIP AT EAVES

For the final example of irregular roofs, a construction is considered combining

1. unequal pitches,
2. an oblique hipped end, and
3. side wall plates out of parallel.

Item 3 poses the greater problem which must be solved either by:

(i) making the roof slopes winding. This would only be possible beyond mimimum limits if the roof were covered say by sheet metal or felt;

(ii) having an inclined ridge, not acceptable to many people;

(iii) raking one set of eaves upwards to the narrow end;

(iv) having a tapered level flat covered with metal flashing at ridge level. This, in the author's opinion, is the most likely solution and will be dealt with later.

The drawing in Fig. 347 shows a roof to the plan given, with two slopes at 45 deg., one at 30 deg. and a gable end.

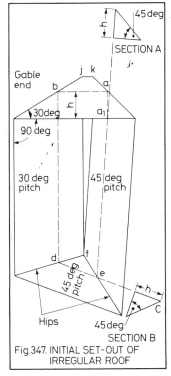

Fig.347. INITIAL SET-OUT OF IRREGULAR ROOF

Some thought is necessary to set out the plan, other than by trial and error. As the gable end is at 90 deg. to the 30 deg. surface, this can be the pitch of the gable wall; but as the 45 deg. slope is at an angle, the gable pitch will be slightly different. It may be obtained as follows: Section A square to the related wall plate is set up at a taken height of 'h' and its inner limit projected to the edge of the plan as shown. A vertical from this point to height 'h' at point 'a' gives the angle for drawing the gable. Section B is also set up to the same height and square to the hipped end, giving point 'c'.

Lines from 'b', 'a$_i$' and 'c' drawn parallel to the respective wall plates in plan intersect at points 'd' and 'e' through which the hips may be drawn to intersect at 'f'. From 'f', lines drawn parallel to the wall plates give the ridge outlines of the side slopes and delineate the narrow tapered flat. Projection upwards on the gable end should give intersecting points 'j' and 'k'. Line 'jk' should be level.

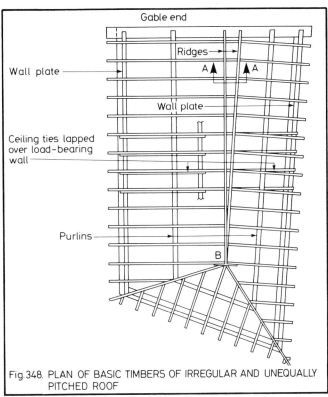

Fig.348. PLAN OF BASIC TIMBERS OF IRREGULAR AND UNEQUALLY PITCHED ROOF

The geometry already discussed may be used to obtain the necessary lengths and bevels, but there are some special practical problems.

First, unless we are prepared to splay all the top edges of one or two sets of rafters, all the rafters must be square on plan to their respective wall plates. This means that the rafters on one side cannot be in the same vertical plane as the opposing ones on the other side. This is shown in plan in Fig. 348. Thus a pair of rafters cannot be directly tied to ceiling joists to give parallel fixings for, say, plaster board ceilings.

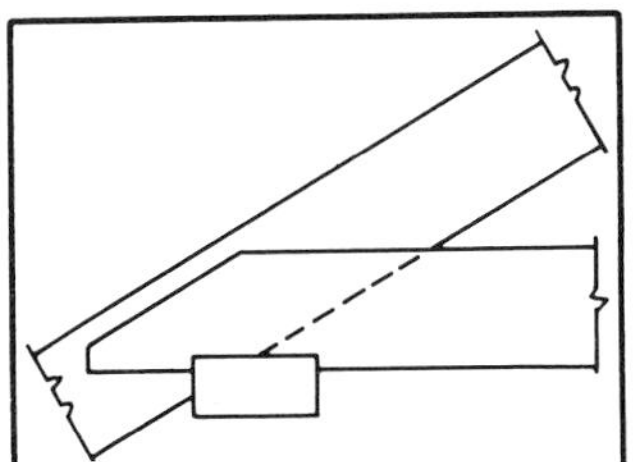

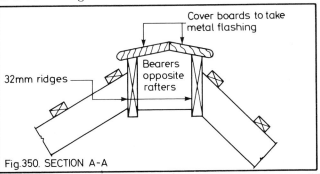

Fig. 349. CEILING TIE FIXING

In the example given, the ceiling joists are laid in two lengths lapped over a load-bearing wall and are nailed to the rafters only on the left-hand side. A few only are shown for the sake of general clarity. To enable them still to act effectively as ties, they may be notched over the wall plate as shown in Fig. 349.

Fig.350. SECTION A-A

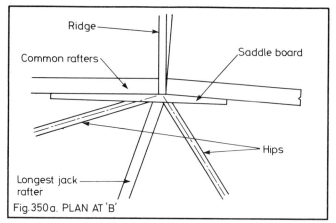

Fig. 350a. PLAN AT 'B'

As the paired rafters are not in common vertical planes, they cannot be easily cleated together at the top. Instead of this, two ridge boards are first made up into a kind of tapered ladder with cross bearers to take the cover board as shown in Fig. 350. This is made up and nailed between the tops of the rafters. The hips fit against a saddle board tapered on plan, to fit the end rafters as shown in Fig. 350a.

FLAT ROOF CONSTRUCTION

One disadvantage of flat roof construction is that there is no or very little air space between the exposed surface of the roof and ceiling beneath, with the consequence that, unless special provision is made there will be a rapid loss of heat; and also the risk, in the case of hollow construction, that warm humid air from the occupied room below will pass through the porous plaster ceiling and strike the cold surface of the upper roof sheeting causing considerable condensation. Hollow timber flat roofs should therefore always be ventilated to reduce the risk of a moisture build up.

The Building Regulations require that the U value of ceilings and roofs combined shall not exceed 0.6 W/m² deg. C. These values can be worked out for any construction if the surface resistance, conductivity of each layer, and widths of air cavities are known but the regulations give deemed to satisfy examples covering a fairly wide range of constructions.

The insulation may consist of various cellular materials in loose or sheet form, fixed anywhere within the roof space, with or without a reflective aluminium foil immediately above the ceiling with its polished side to the cavity.

In my opinion, the best treatment is to lay a suitable thickness of, say, insulation board on the joists directly under the roof boarding or, if practicable, under the roof covering, and to nail aluminium foil or foil-faced plaster board to the joist above (or to form), the ceiling.

The foil being reflective will reduce the loss of heat by radiation and will also form a satisfactory vapour barrier against moist air from the room below, while the insulation board will present a warm surface to any humid air within the roof and reduce the risk of condensation. As these details do not influence general construction details, they have been left out in the illustrations.

Coverings to flat roofs must be impervious to water and the joints are either sealed or lifted above the general level of water flow. It is not good, however, to have water standing on a roof. Rain water, particularly in towns, may

contain weak acids likely to have a deleterious effect on metal roofing when permitted the continuous contact in standing pools. The fall of any part of a roof should be sufficient to ensure that pools cannot form. The minimum fall is usually given as 1 : 96 for metal roofing and 1 : 60 for built-up bituminised felt roofs. Any gutters should have a similar fall and be of such dimension as to hold storm water without undue build-up. When a building is in an exposed position, it is better to increase the fall.

When the roof is sheeted with timber boarding, some inevitable moisture movement in the timber may cause cupping (Fig. 351). With a fall of 1 : 96, a 150 mm board only needs to lift $1/96 \times 75 = 0.75$ mm to cancel out the fall when they are laid across the fall. Uneven cupping of the boards is also likely to cause uneven wear in the covering. For this reason, if there is a choice, say with square edged boards, flat sawn boards should be laid heart side up (see Fig. 352).

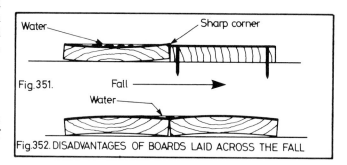

Fig. 351.   Fall

Fig. 352. DISADVANTAGES OF BOARDS LAID ACROSS THE FALL

If roofing boards are laid with edges parallel to the fall, cupping of the boards will not affect the water flow, but if joists are to span the short way, which is structurally advisable, the flow must be the long way with the increased risk of build-up of rain water at the eaves in a heavy storm. One way out of the dilemma is to use diagonal boarding; in which case noggings must be cut between joists to support the splayed ends of the boards.

The overall design of a flat roof depends upon the type of impervious covering used. If the roof is to be felted or asphalted, a perfectly flat roof, laid to the correct fall, is all that is needed. Asphalt to timber roofs should be laid on felt to permit movement of the timber under the asphalt. Against a wall, the asphalt should be turned up against a wood skirting (Fig. 353) with a 12 mm gap eventually covered by lead flashing.

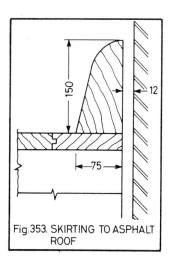

Fig. 353. SKIRTING TO ASPHALT ROOF

Boarding materials

Various kinds of boarding materials may be used for roofing materials when areas are not limited by drips and rolls necessary to sheet metal roofing. The most common of these are as follows:

PLYWOOD: This is the obvious alternative to solid timber. It has the advantage of being in large sheets and should be nailed to the joists in the usual way. In view of its lack of stiffness, as compared with timber, all the joints not resting on joists should be supported by nogging pieces. It should be nailed at 150 mm centres on edges and intermediate joists. The thickness of the plywood depends upon the spacing of the joists, type of plywood and anticipated dead and imposed loading.

WOOD WOOL: This is laid with its length across the joists and should .be at 600 mm centres .for 50 mm .slabs. For other thicknesses and constructions it is better to consult the manufacturers. It is nailed with special large-headed galvanised nails using 5 nails across the width of the slab to every joist. The edges should be covered or protected and a 12 mm cement screed (1:4) should be laid over the roof.

COMPRESSED STRAW SLAB: This is usually 50 mm thick in sizes of 2.4 m by 1.2 m and should be laid with its length parallel to the joists at 600 mm centres. All end or heading joints should be supported by noggings. Special types with bituminised shower proof facings should be used for roofing work. It should be nailed with 100 mm large headed galvanised nails at 225 mm centres. All edges exposed by cutting should be sealed with special sealing strips. Joints between slabs should be sealed with weathering scrim.

CHIPBOARD, OR PARTICLE BOARD: This is made from wood chips with a resin adhesive and it should be treated in the same way as plywood but needs closer supports as it lacks some of the stiffness of plywood. It should be protected from the weather until finally covered.

Finishing flat roofs

Flat roofs may be finished either overhanging the external walls of a building, flush with them, or within parapet walls. Figs. 354 and 355 show section and part plan of a roof boarded across the fall and suitable for a built-up bituminised or asphalt finish. The joists are built into the wall at the back and are carried on 100 mm by 50 mm wall plates at the front. Sprocket pieces in line with the fall of the roof are nailed to the joists to give extended support out to the fascia to provide a satisfactory drip and suitable ventilation to all the air spaces within the joists.

Short sprocket pieces are also nailed at right-angles to the end joists and supported on wall plates on the outside of the end walls. A continuous bearer nailed to the soffit and notched into the sprockets, carries the ends of the roof boarding. A diagonal sprocket carries the mitred intersection between verge and eaves.

Figs. 356 and 357 show a section and part plan of alternative constructions with joists laid across the fall. Instead of tapered firrings, there are parallel strips of progressively increasing widths nailed to the tops of the joists to give the necessary fall to each board. Otherwise construction is generally the same as before. The ends of the boards are this time carried by a bearer behind the fascia.

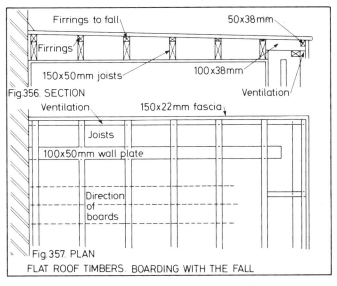

Fig.356. SECTION

Fig.357. PLAN
FLAT ROOF TIMBERS. BOARDING WITH THE FALL

As already stated, the limited fall to a flat roof, which is an essential part of its design, leaves very little room for inaccuracy in construction. Variations in sawn widths of joists, of the same nominal size, or a slight taper end to end may be sufficient to give a fall the wrong way. These

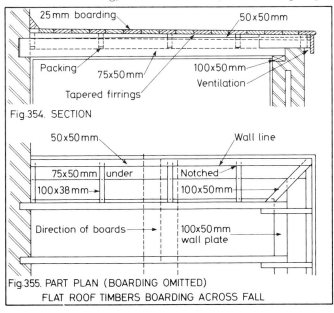

Fig.354. SECTION

Fig.355. PART PLAN (BOARDING OMITTED)
FLAT ROOF TIMBERS BOARDING ACROSS FALL

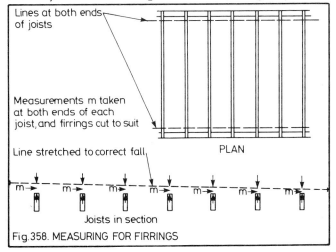

Fig.358. MEASURING FOR FIRRINGS

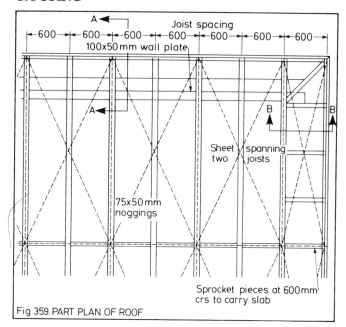

Fig.359. PART PLAN OF ROOF

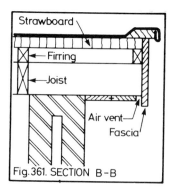

Fig.361. SECTION B-B

errors can be compensated for if a line to the actual fall, or the required level as the case may be, is stretched across each end of a row of joists. Measurements taken from the line to the tops of the joists give the separate end to end width of each firring, as shown in Fig. 358.

The part plan of the timbers to an overhanging flat roof to receive 2,400 mm by 1,200 mm compressed straw slabs is shown in Fig. 359. The crossed dotted lines delineate the ultimate positions of the slabs. The fall of the roof is parallel to the joists. Nogging pieces are fixed under each heading joint in the slabs. Sprocket pieces carry the narrow make up slab to the end of the roof. These are nailed to the first joist and to the fascia; while at the end of each slab the sprocket is carried through to the next joist, thus increasing the length of the cantilever support. This stiffening member and the other sprockets rest on the external wall.

Figs. 360 and 361 are sections through the roof at A-A and B-B in plan and show the finish to the eaves and verge. From Fig. 360, it will be seen that the end of the joist takes the fascia with a wood fillet to form a drip support. A continuous cleat nailed to the fascia and notched to the joists, supports the edge of the slabs. The soffit boards are nailed to the underside of the extended joists and are a little narrow to provide ventilation. The gap should not be sufficient to allow the entry of birds. The joists are carried by a wall plate.

Built-up felt roofing is shown but the construction is

also suitable for asphalt. Tapered firring strips give the fall to the roof.

Fig. 361 shows a section through the verge with the short sprocket pieces and wood capping to lift the felt and provide a check against flow of water over the verge. (Figs. 362 and 363 show alternative constructions where there is just sufficient overhang to allow ventilation. The depth of the joist ends has been reduced to give a narrower fascia without exposing the joist ends.

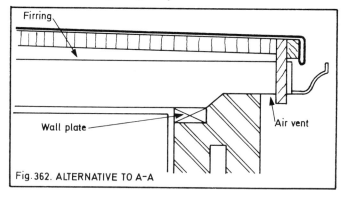

Fig.362. ALTERNATIVE TO A-A

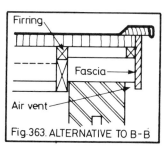

Fig.363. ALTERNATIVE TO B-B

Metal roof coverings

When the roof is covered in metal, such as lead, copper or aluminium, there is the problem of temperature movement from summer to winter. To deal with this two precautions are taken:

1. Every sheet is kept to a limited area with maximum permitted widths and lengths.
2. Individual sheets are generally fixed on one side and one end only the other edges being captive but free to move.

Two types of joint are needed: one parallel to the fall which stands well above the roof and may be self-supporting or dressed around a wood roll; the other across the fall must not be upstanding or it would trap the water. In the case of copper and aluminium sheeting, which is comparatively thin, the cross joints are flat welts which are too

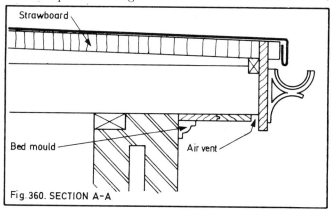

Fig.360. SECTION A-A

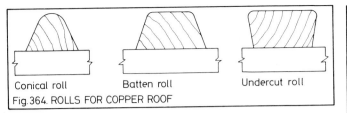

Fig.364. ROLLS FOR COPPER ROOF

Conical roll | Batten roll | Undercut roll

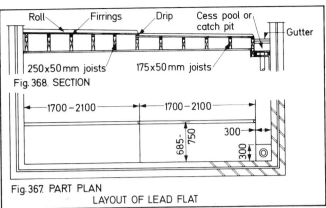

Fig. 368. SECTION

Fig. 367. PART PLAN
LAYOUT OF LEAD FLAT

thin to form an appreciable check. The rolls parallel to the fall are one of those shown in Fig. 364. The sheets are secured to the rolls by means of cleats nailed to the rolls. They are incorporated in the welted joints to the covering caps holding the sheets on either side without full fixing.

Lead sheeting is also secured by rolls which are spaced at from 680 mm to 750 mm centres. Each sheet is dressed over the roll to form an undercloak and securely nailed; it is then dressed over the undercloak of the adjoining sheet, without nailing, to form an overcloak. Fig. 365 gives details of the lead roll which is undercut to give a better retention of the overcloak.

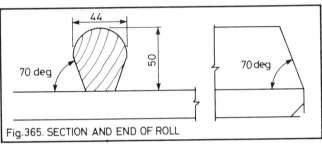

Fig.365. SECTION AND END OF ROLL

The joint across the fall cannot be formed with a flat welt as the lead is too thick and therefore a step-down, known as a drip, is formed across the fall of the roof at intervals of from 1,700 mm to 2,100 mm (Fig. 367). The lower sheet is dressed up over the drip into a rebate to accommodate its thickness and securely nailed. The overcloak from the sheet above dresses straight down over it. A section through the drip is shown in Fig. 366. The 'V' joint shown enables a groove to be formed between the undercloak and overcloak as a check to capillary action.

Figs. 367 and 368 show part plan and section through

the lead flat prepared for the plumber. The roof in this case is contained within parapet walls. This poses a problem for the disposal of rainwater. The gutter is contained within the roof area and is formed in lead fitted to boxing provided by the carpenter. The gutter has to have falls and drips along its length of the same order as is needed in the length of the roof. If joists are across the fall, the last joist supports, and indeed forms, the inside of the gutter. The bottom is carried by cleats nailed to the joist and holding cross bearers which rest at their outer ends on similar cleats nailed to the wall.

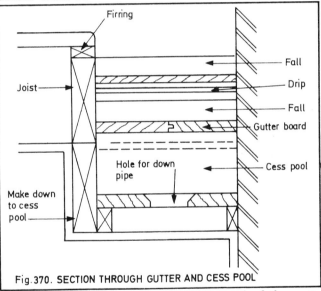

Fig.370. SECTION THROUGH GUTTER AND CESS POOL

The water is collected at the lower end of the gutter in a cess pool or catch pit, which is a lead-lined square box holed to take the down pipe to a convenient drain or soakaway. Fig. 369 is a longitudinal section through the gutter and cess pool, showing the cleats nailed to the joist and carrying the gutter bearers. In this case, a partition wall halves the span and enables normal sized joists to be used. The lower half of the gutter and cess pool are below ceiling level and a double depth joist is used to contain them. Fig. 370 is a section through the gutter and cess pool to a larger scale.

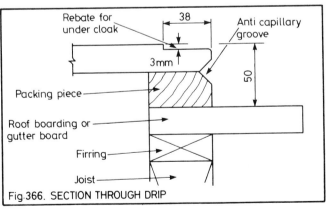

Fig.366. SECTION THROUGH DRIP

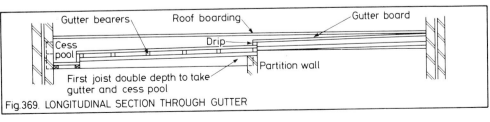

Fig.369. LONGITUDINAL SECTION THROUGH GUTTER

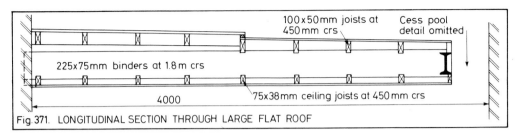

Fig. 371. LONGITUDINAL SECTION THROUGH LARGE FLAT ROOF

Larger flat roofs

When a flat roof has to be formed to a hall or large room where no intermediate supports are possible, the technique used in the construction of double and framed floors becomes necessary. Some ingenuity may be required in order to limit the amount of material needed to deal with the accumulated depth of several drips and rolls.

A section through a flat roof spanning 4,000 mm the short way, but being of much greater length, is shown in Fig. 371. The roof and ceiling are carried by 225 mm by 75 mm binders at 1.8 m centres. The joists are notched over the binders to give the necessary fall to the lower end while the fall at the upper end is given by tapered firrings. The ceiling joists are also notched over the binders and are of suitable size to carry 9 mm plasterboard at 1.8 m centres. The whole of the weight of the lower half of the roof is carried by the steel joist, the size of which depends upon its span.

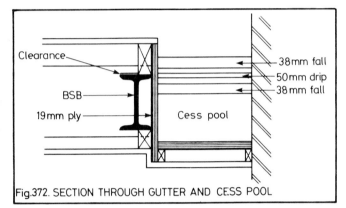

Fig.372. SECTION THROUGH GUTTER AND CESS POOL

Fig. 372 is a section through the cess pool and gutter using plywood to form the casings. The ceiling is also stepped down and carried through.

CHAPTER 7

Stair Construction

The construction of timber stairs, especially of the more elaborate type, has always been regarded as the cream of joiner's work to be given to selected craftsmen in the shop. When, as is usual in public and industrial buildings, the stairs are of concrete, it is still the carpenter who makes and fixes the formwork and whose precision and knowledge of requirements to hold and support the wet concrete are largely responsible for the success of the job.

The casting of the stairs is usually carried out as a separate operation from that of the main structure so as not to delay the main structural work, a key being left in the face of the edge beam or trimmer to the floor to take the stairs. It is better that this beam is kept about 75 mm back to allow some adjustment in the setting up of the stair forms.

FORMWORK FOR CONCRETE STAIRS

Before starting work on the formwork, all relevant information should be obtained. This may include thickness of facings or finishings, such as rubber, terrazzo or granolithic, and any special provisions to be made for fixings, such as balusters. It may require provisions for bolts or dovetail fixing fillets to be cast in or sockets to be formed. Going and rise of the stairs must suit site conditions and will seldom be precisely as the scale drawings. The position of the finished floor level must be checked off the datum line if these floors are not down.

An exact location of the first riser must be obtained and the proposed face of the landing riser plumbed down to give the stair going. The precise rise must then be confirmed by levelling back from the first riser to a point plumb under the landing riser measured up.

The construction of stairs is governed by the 1985 Building Regulations "Stairways and Ramps Detailed in Approved Document "Stairways, Ramps and Guards". Reference to it within this chapter are confined to the problem it imposes on setting out.

Stairs have been divided into four groups identified in table

1 "Rise and going. Fig. 373 'a' to 'd' gives practical methods of obtaining step sizes and proportions according to which heading the stairs come under.

Obtaining step sizes and proportions

Two rules common to steps for all staircases are:

1. Twice the rise of a step plus the going must not be less than 550 mm and not more than 700 mm and
2. the minimum rise must be 75 mm.

Other regulations vary according to the heading the stairs come under.

In the graphs, the going is measured along the base line from 0 (see 'a'). Each graph is set up to the extremes in rule (1) above and then individually modified to suit a particular case. Thus, if the rise is 0 mm and the tread is measured from 0, it will be 550 mm minimum and 700 mm maximum. Similarly, if the going is 0, then rise may be from 275 to 350 mm (700 and 550 mm divided by two).

If these points marked on the graph are joined up and the actual going drawn from point 0, the rise can be to any point between the raking lines. Further limiting regulations are as follows.

CASE ONE: Fig. 373 'a', pitch must be less than 42 deg. so the riser does not come into the shaded area to the left of the 42 deg. pitch line. The going must not be less than 220 mm so the riser must not come to the left of the relative vertical line.

The rise must not be more than 220 mm so the riser must finish under the relative horizontal line. The rise must not be less than 75 mm, so the riser may not come into the shaded area on the right or finish below the dotted line.

CASE TWO: Fig. 373 'b' is the graph for case two for which maximum rise is 190 mm, minimum going 240 mm and maximum pitch 38 deg.

CASE THREE: Fig. 373 'c' illustrates case three where the minimum going is 280 mm and maximum rise 180 mm. The pitch is not identified but cannot be more than 32.70 deg.

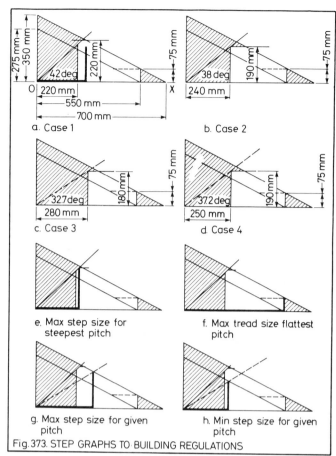

a. Case 1   b. Case 2

c. Case 3   d. Case 4

e. Max step size for steepest pitch   f. Max tread size flattest pitch

g. Max step size for given pitch   h. Min step size for given pitch

Fig. 373. STEP GRAPHS TO BUILDING REGULATIONS

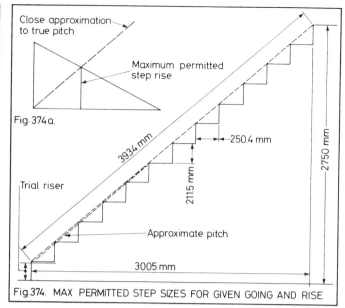

Fig. 374a.

Fig.374. MAX PERMITTED STEP SIZES FOR GIVEN GOING AND RISE

CASE FOUR: In Fig. 373 'd' which illustrates case four, going must not be less than 250 mm and rise not more than 190 mm giving a maximum pitch of 37.20 deg.

Fig. 373 'e' to 'h' indicate adjustments to step sizes within the confines of the relevant graph.

Although case one has been used each time, the same possibilities of adjustment exist for the other three. Thus Fig. 373 'e' illustrates the maximum pitch for the largest possible step where the minimum number of steps are needed.

In Fig. 373 'f', the widest tread and the shallowest rise are indicated. Fig. 373 'g' shows the maximum step size when the pitch is given; and Fig. 373 'h' the smallest for the same pitch.

It is often necessary to fit stairs into the smallest permitted space with as few steps as possible. The procedure for doing this, illustrated in Fig. 374, is as follows. Assuming that the stair going is 3005 mm and the rise 2750 mm, the first step is to set up an approximate pitch line by drawing going and rise to scale, setting up an approximate riser, say, 200 mm, and drawing in the pitch. Parallel to this, a step graph (maximum line only) is constructed and put in the pitch line, as in Fig. 374 'a', giving the maximum step rise, measured at, say, 215 mm.

$$\text{Minimum number of permitted risers} = \frac{2750}{215} + 1 \text{ for}$$

remainder = 13 risers. Step rise = $\frac{2750}{13}$ = 211.50 mm.

There will be only 12 treads, so step going

$$= \frac{3005}{12} = 250.40 \text{ mm}.$$

Set out of string

As site conditions, particularly for formwork, are seldom conducive to fine work, it is better to calculate the nosing line length and divide this into the required number of steps rather than risk repetitive errors. Note that the nosing line comes from the first riser, *not* the floor; so that the rise of the nosing line = 2750 − 211.50 mm = 2538.50 mm and (by Pythagoras' rule) length of nosing line

$$= \sqrt{3005^2 + 2538.5^2} = 3934 \text{ mm}$$

and hypotenuse to one step = $\frac{3934}{12}$ = 327.8 mm. (note that the one decimal place of the millimetre permits a very accurate final estimate).

A half turn stairway

Fig. 375 shows the plan of a stairway which is to have half turn open well stairs with steps between the quarter

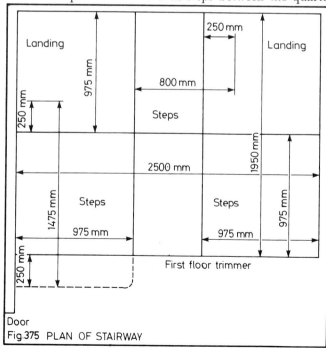

Fig.375 PLAN OF STAIRWAY

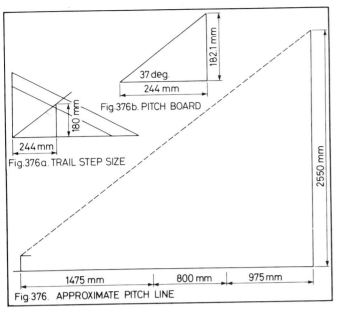

Fig.376b. PITCH BOARD

Fig.376a. TRAIL STEP SIZE

Fig.376. APPROXIMATE PITCH LINE

space landings, the overall rise being 2550 mm. It is assumed that it must conform to the Building Regulations in case two, i.e. maximum permitted step rise 190 mm, minimum going 240 mm and maximum pitch 38 deg.

It is obvious that two steps will have to be placed between the quarter space landings. An approximate pitch line is obtained by adding together the goings of the three flights, including an estimated step width for each landing, and setting these against the total rise with an estimated first riser as before. The step graph as in Fig. 376 'a' will give a choice of step sizes.

From the first floor trimmer to landing (Fig. 375) is 975 mm which presumably cannot be altered and will not take more than four steps. Therefore trial tread

$$= \frac{975}{4} = 243.75,$$

say 244 mm which, on the graph, gives a trial rise of 180 mm (below maximum).

$$\text{Number of risers} = \frac{2550}{180} = 14.17.$$

Therefore try 14 risers.

$$\text{Step rise} = \frac{2550}{14} = 182.1 \text{ mm},$$

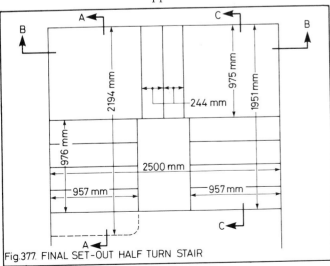

Fig.377. FINAL SET-OUT HALF TURN STAIR

so the step proportions conform to the regulations (Fig. 376 'b'). The set out of the plan may now be completed as in Fig. 377. Note that, as all the steps have to be the same size, those between the quarter space landing come inside the well lines.

Erecting the formwork

The formwork to these stairs will next be considered as an example of straightforward construction. The stairs themselves are basically raking slabs; and the formwork to support the soffits must be strutted and braced back to the rear wall for the final thrust. The decking may be of solid timber with loose tongues, or covered in hardboard, or of proprietary steel, or framed plywood panels and supported by either horizontal joists, raking ledgers or raking joists on horizontal ledgers. The former gives a ready method of supporting the outer string.

Fig. 378 shows section 'A-A' through the formwork to the lower part of the stairs. The low heights do not warrant the use of a proprietary propping system. The ledgers on the landing forms take the thrust of the raking ledgers to the stairs, the whole system being braced both ways.

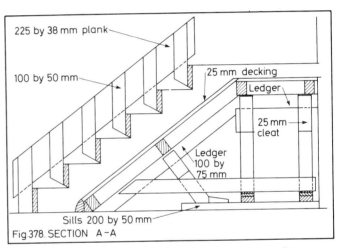

Fig 378. SECTION A-A

Fig. 379 section 'B-B', shows a section of the formwork to the quarter space landings and the steps between and shows how the landings also support the raking ledgers to the intermediate steps.

Fig. 380, which is section 'C-C' on the plan, shows the

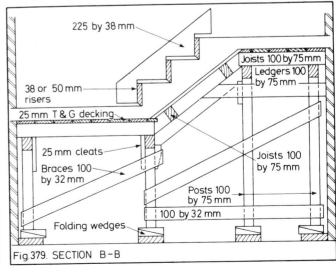

Fig.379. SECTION B-B

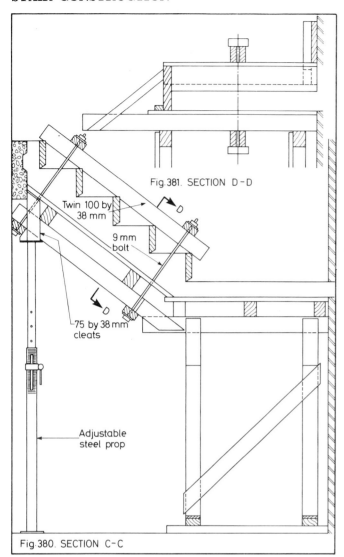

Fig. 380. SECTION C-C

Fig. 381. SECTION D-D

Twin 100 by 38 mm

9 mm bolt

75 by 38 mm cleats

Adjustable steel prop

top end of the upper flight supported by adjustable steel props. Note the block inserted to give a level bearing on the prop head plate and the cleats nailed to the ledgers and blocks which hold the formwork against the trimmer.

Where the step risers abut a wall, they may be carried on hangers nailed to a plank which is nailed to the wall as in Fig. 378, or on a notched 38 mm plank as in Fig. 379.

The outer string, which is notched to take the risers, is supported by ribbons nailed to the decking and struts taken down to the joists as shown in Fig. 381. The tendency for the risers to bulge under pressure of wet concrete is countered by a runner bolted through the concrete or strutted off the ceiling, whichever is the more convenient, as shown in Figs. 380 and 381.

Striking of stair forms

Once the stairs are concreted into position, care should be taken to protect them against ill use, particularly on the arrises and nosings. Striking should be delayed until the concrete has attained sufficient strength. In the interest of safety, the stairs should be fully guarded before being used.

If the surface of the step is to be fair face finished, then the riser boards should be bevelled underneath to permit trowelling of the concrete over the whole surface of the step. The bottom edge of the riser being left flat helps to

prevent upsurge. Fig. 382 shows a plywood outer string stiffened by twin runners and not notched to shape. This makes the filling and screeding off more difficult, but is satisfactory for steps which have afterwards to be faced.

Steps formed with moulded nosings are best constructed by building up from the riser board to obtain the shape as in Figs. 383 and 384.

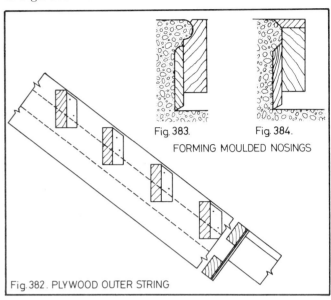

Fig. 383.   Fig. 384.

FORMING MOULDED NOSINGS

Fig. 382. PLYWOOD OUTER STRING

TIMBER STAIRCASES

As has already been stated, the construction of timber stairs is traditionally a job for the workshop; it is therefore outside the scope of this book which is generally confined to site operations. The preliminary work of measuring up and the final fixing on site, however, are equally important parts of the work. Without their successful execution the stairs, no matter how well made, would be doomed to failure. The following example deals with some of the problems which may be encountered on site; while constructional details, which are normally prepared in the shop, are left out.

Fig. 385 is a plan of a half turn open newel staircase starting with a bullnose step to the bottom flight, quarter space winders (tapered treads in the Building Regulations), a second flight to a quarter space landing leading to a mezzanine floor to the bathroom, and finally three steps to the first floor. Windows and doors have been included to give a more complete example of what is likely to be expected.

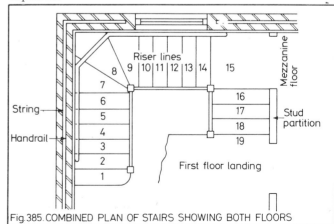

Riser lines

Mezzanine floor

String

Handrail

Stud partition

First floor landing

Fig. 385. COMBINED PLAN OF STAIRS SHOWING BOTH FLOORS

Although we get a picture of the general requirements of the job from the plan and perhaps various sections and elevations in the architect's drawings, it is never safe to assume that given measurements have been followed with the required degree of accuracy. There may have been last minute changes not shown on the original plan. Other difficulties, not evident from the drawings, are likely to be met on the site. It is therefore safer to work to dimensions taken in situ.

Preparations on site

Assuming that the reader is responsible for these site measurements, the following recommendations should contribute to his confidence and the ultimate success of the job. Avoid making sketches on loose paper as these can get lost. Use a good sized flat notebook with stiff covers to provide flat drawing surfaces (squared paper or alternate squares and lines are convenient). Form the habit of making good-sized ruled sketches approximately to scale and in good proportion. This may require some thoughtful work and perhaps some initial corrections with an eraser.

Before starting on the work, give some consideration to the job. Study the drawings and specifications with a view to what problems they are likely to introduce.

Prior to making any measurements, complete all the sketches necessary to provide the information required and insert all the dimension lines with arrow heads extending to the exact limits of the measurement to be indicated. Measurements can then be written down as they are taken; thus reducing the possibility of error. Make sure that all sizes are correct, measuring twice if in doubt. It is also advisable to check with the architect's drawings, investigating any appreciable discrepancy.

The measurements required for the stairs in Fig. 385 are shown in the sketches in Figs. 386 to 389. The head of the page should be dated and the job fully identified.

Make a plan of each floor level (Fig. 386) checking the sizes between walls on each plan and indicating whether the plans are vertically over each other. There may be differences due to offsets in the brickwork at a floor level.

State whether the measurements are to bare brickwork or to face of plaster. If the former is true, indicate the thickness of the plaster or fixing grounds. It may be that the strings have to continue the line of the skirting; in this case, the thickness of the plaster will influence actual string thickness. Show the direction and relative positions of joists on each floor.

Make an elevation with sections of each wall (Figs. 387 to 389) showing the height and width of openings. Take careful measurements of vertical heights of various floors and landings. When possible, these should be marked direct on a storey rod (Fig. 390) which can be sent to the shop as a check for future dimensions. Test the walls for plumb and the corners on the plan for square, and record any unusual variations.

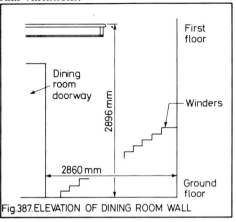

Fig.387. ELEVATION OF DINING ROOM WALL

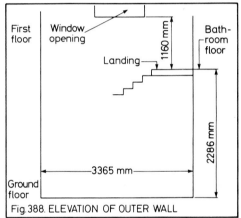

Fig.388. ELEVATION OF OUTER WALL

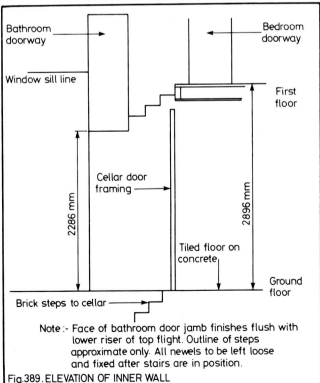

Note :- Face of bathroom door jamb finishes flush with lower riser of top flight. Outline of steps approximate only. All newels to be left loose and fixed after stairs are in position.

Fig.389. ELEVATION OF INNER WALL

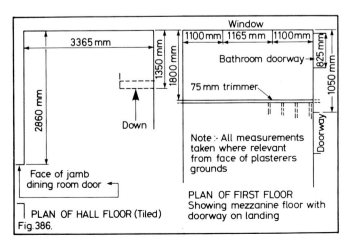

Note :- All measurements taken where relevant from face of plasterers grounds

PLAN OF FIRST FLOOR
Showing mezzanine floor with doorway on landing

PLAN OF HALL FLOOR (Tiled)
Fig.386.

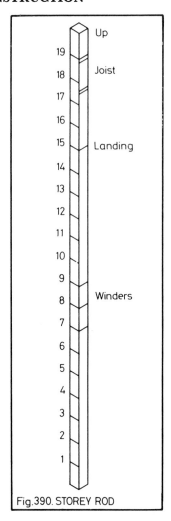

Fig.390. STOREY ROD

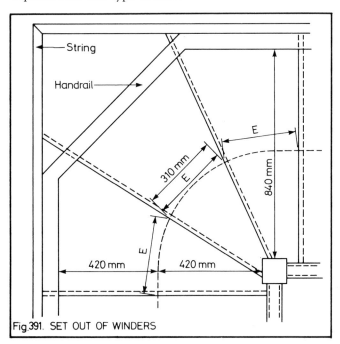

Fig.391. SET OUT OF WINDERS

housed in the newel; for sound construction it is better if only two are housed.

These are placed in the middle of the newel face to give a solidly enclosed housing. The details are shown in Fig. 391 from which should also be noted the method of spacing out the winders on the centre line. They can no longer be at angles of 30 deg. and 60 deg. to the fliers.

The requirement to fix a handrail on the outside of the winders, and that it should be continuous, brings in a new condition. The obvious answer to a need for a continuous handrail would be to provide one which was a quadrant in plan, i.e. wreathed, but this would be unreasonably expensive for this type of stair.

It would be equally unreasonable to take the handrail through to the corner, as this would defeat the object of keeping the pedestrian along the recommended line of travel. The solution given in the plan is economical and satisfactory.

Before actually taking sizes, it is important to survey the area and look for restrictions to access which may limit the size and therefore the degree of assembly of stair units; e.g. it may be necessary to leave all newel posts and balustrades loose.

The space under the stairs may be left open; in which case, a plaster soffit is the usual finish. Alternatively the space may be enclosed to form a cupboard by taking the newel post to the floor and enclosing the area with rectangular and triangular panelling, known as spandrel frames or drags. These are rebated under the strings and housed into the continuous newel posts.

Requirements of the Building Regulations

The Building Regulations 1985 in Approved Document K deal comprehensively with stairways etc. and identifies special requirements for four different groups of buildings. It is necessary that each example given here should comply with the regulations for the group to which it belongs. To this end the regulations are partly explained.

The example given in Fig. 385 comes under building or compartment of purpose group I or III for any stairway within a dwelling or serving exclusively one dwelling. The width, clear of handrail, balustrade and wall (which may extend into the stairs), must be at least 800 mm. Within this width, the minimum going of winders must be 50 mm. The going of winders, measured on the centre of the width, must be uniform and not less than 220 mm. The aggregate of twice rise plus going must be between 550 mm and 700 mm; and a handrail must be provided at the wide end of the winders.

Other regulations are straightforward and it is recommended that the reader refers directly to them.

Complying with the Building Regulations

The traditional way of fitting winders around a newel post, which was to make the face of each riser radiate from the centre of the newel, cannot be used as this will not give sufficient width of step at the narrow end. However there is no reason why all four risers should be

Fixing the staircase in position

It is now assumed that the stairs have been delivered on site and assembled as far as is convenient. They now have to be fixed.

The ground floor is shown as tiles on concrete (Fig. 386) As the tiling to the entrance is to be patterned, it is left until the spandrels are in position. The concrete has been laid. It will be necessary to check the relative floor levels to set out the position of each newel and place blocks of wood of the correct thickness for the newels to stand on.

The lower flight is offered up and the approximate line of the wall string marked on the plastered wall. The rendering can be cut away and the wall plugged with dry timber or proprietary plugs for future fixing of the string. If the floor is level, the spandrels can be used as templets for marking the positions of the plugs.

The respective flights are lifted into position and supported temporarily by loose rakers while the newels are attached. The newels should be left unpinned for as long as possible, in case handrails or other members have to be inserted before pinning.

The landing trimmer is supported by inserting one end into the wall and resting the other end in a slot in the newel. The landing will be fixed and connected with the bathroom floor.

The second flight having been placed, the tongued and grooved joint at the connection of the wall strings is now fixed. The tongue on the lower flight supports the one above.

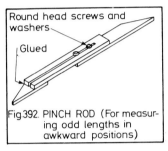

Fig.392. PINCH ROD (For measuring odd lengths in awkward positions)

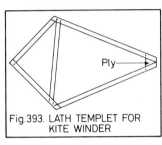

Fig.393. LATH TEMPLET FOR KITE WINDER

The winders should be checked for size but left uncut. The lengths and widths should be checked by means of a pinch rod and templet (Figs. 292 and 293). The pinch rod is very useful for measuring and transferring internal dimensions. The templet will indicate whether the tread can be easily inserted. Some difficulty may be experienced with the middle 'kite' but this may be overcome by easing at the corners.

Each of the three winders may now be placed and the stairs can then be fixed by nailing the wall strings into the plugs. The treads and risers of the winders are screwed, wedged and glue blocked as the fliers have already been done. The bottom step and newel should be ready for fixing although it may be necessary to scribe the bottom-riser to the floor in some cases.

The handrails over the balustrades are tenoned and housed into the newel, the top or bottom of the mould being scribed in square to avoid undercutting housings.

The top flight can be fixed independently of the others after the quarter space landing is in position. Its relative newels with apron lining and nosing to the landing trimmer and the landing balustrade will be easy to assemble.

If there is a capping to the string, this should now be fixed followed by the fixing of balusters. Care should be taken to see that they are correctly spaced and upright. The handrail over the outside of the winders is secured by brackets to the wall as a separate operation. It is advisable to fix this in position and then remove it until the finishing stages of the contract. To avoid risk of damage to the steps during the making good by plasterers and other trades it is usual to fix temporary casings to them.

If all the measurements have been correctly made and worked to, the various parts should come together freely. In practice however, unforeseen difficulties often arise. Corners may be out of square, brickwork uneven, levels slightly wrong or walls not plumb. Problems like this must be dealt with as they arise, and this involves some compromise between true precision and existing conditions.

Fitting carriages to stairs

Stairs, particularly wide ones, need a centre support between strings to prevent the creaking of treads where they get the most wear; although this is often omitted in the interests of economy. The supports are known as carriages and consist of timbers fitted lengthways under the line of steps.

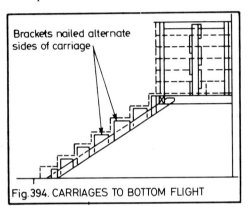

Fig.394. CARRIAGES TO BOTTOM FLIGHT

The support is taken direct to each tread by means of brackets fitted tight against the underside of the tread and nailed to the carriage on alternate sides. These details are shown in Figs. 394 and 395. Under straight flights, carriages are simple to fix; but under winders, where head room is limited, some ingenuity is needed to get the necessary supports and fixings. Figs 395 and 396 show one method with raking timbers using brackets nailed to the carriages under the winders.

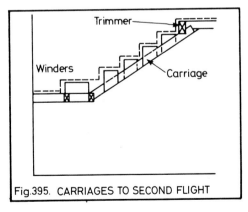

Fig.395. CARRIAGES TO SECOND FLIGHT

Another method is to fix bearers under each winder parallel to the riser, the ends being attached to the newel and wall string. When the soffit or bulkhead is to be plastered, other timbers known as rough strings are fixed inside the strings to act as bearers for the laths.

The method shown allows the bulkhead to be plastered in a series of flat surfaces. When the second method is adopted, the plastered surface will be twisted or flued.

95

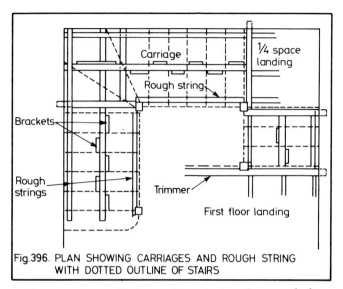

Fig.396. PLAN SHOWING CARRIAGES AND ROUGH STRING WITH DOTTED OUTLINE OF STAIRS

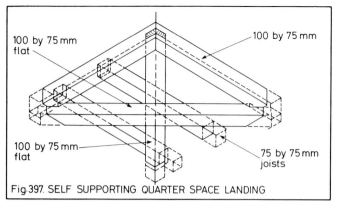

Fig.397. SELF SUPPORTING QUARTER SPACE LANDING

Circumstances will determine whether the spandrels or carriages are fixed first. In the present case, the spandrel might be inserted directly after the newels are pinned.

Fitting the landings

A half space self-supporting landing to a half turn open newel or dog leg stairs presents no problem. The trimmer spans between two walls and takes the ends of short joists notched into it. If no support is given by the newel post, the size of the trimmer must be adjusted to the increased span.

Where the stairway is 'L' shaped (as described in Chapter 3—Timber Floor Construction), the corner projecting into the opening is supported by cantilever joists as shown. Flights of stairs taken straight from the floor to a quarter space landing themselves tend to act as props to the landing.

When, however, both ends of a flight are supported by intermediate landings, it is better if they are both self-supporting. In any case, subsequent movement due to settlement or shrinkage may lead to creaking. In Fig. 397, a diagonal bearer projecting from the internal corner of

the wall is supported midway by another at right angles with both ends bedded into the wall. The end of this cantilever carries the trimmers which, with the joints, complete the landing assembly.

To reduce the amount by which the main diagonal bearer stands below the landing framing, this and the cantilever consist of 100 mm by 75 mm timbers laid flat, the joists being 75 mm by 75 mm and the trimmers 100 mm by 75 mm fixed with the width vertical.

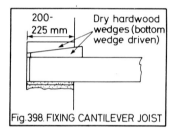

Fig.398. FIXING CANTILEVER JOIST

If trimmers can be bedded 200 mm to 225 mm into the wall, they will act satisfactorily as cantilevers. They should be dry timbers because, a slight shrinkage will cause them to drop. A level bearing under the cantilever must be ensured: if necessary with a mortar bed reinforced with a strip of expanded metal. It should be tightened down with folding wedges, as shown in Fig. 398, the bottom wedge being driven over the timber and not against the brickwork.

CHAPTER 8

Partitions

The purpose of partitions is considered here as being to divide a particular storey into separate rooms within the same occupancy. The Building Regulations regarding heat and sound insulation do not therefore apply.

The popular use of trussed rafter and cross wall construction enable the dead and imposed loads from roof and floor to be transmitted directly to the main walls, so partitions themselves do not have to carry any imposed loads and need only be strong enough to provide the necessary rigidity and resistance to horizontal impact. They do, however, add to the weight of the floor and should be as light as possible for this reason.

A number of different types of non-loadbearing partitions are available involving the partial or full use of proprietary materials. They may be of cellular plasterboard construction, or woodwool slabs stiffened and supported by wood or metal framing, or may be of hollow metal with insulation infill. This chapter is confined to those partitions in which timber is the main structural element.

In earlier designs, timber partitions were sometimes framed up as trusses and, besides their own weight, had to carry the floor at sill level and the ceiling at head level. The presence of a door opening, generally to one side, interfered with the planning of the strut position. Effective strutting was really confined to the area of wall above door height or intertie level. With the low ceiling heights permitted by the Building Regulations for domestic work, this is not wide enough to be effective and other methods of support are more satisfactory. The use of timber is therefore confined to framing cut between floor and ceiling, resting on the floor and faced with suitable lining materials.

COMMON OR QUARTERED PARTITIONS

The simplest type of partition, known as common or quartered, consists of a head nailed to ceiling joists or to bearers cut between the joists, a sill nailed to the floor, vertical studs cut between, suitably stiffened, and faced with lining materials.

The thickness of each stud should be sufficient to give a satisfactory nailing to sheets meeting in the middle with a reasonable allowance for inaccuracy, say 50 mm. The overall dimensions governing the ultimate strength should be sufficient to give the required stiffness against reasonable impact and will be greater as the height increases. For a ceiling of, say, 2 m height, 75 mm by 50 mm studding should prove satisfactory; while for a height of 2.5 m, 100 m by 50 mm studding will be needed.

Fig. 399 is the elevation of a stud partition with a doorway before the linings and finishings have been applied. The spacing of the studs depends on the superficial dimensions and thickness of the type of lining used. If lathed and plastered in situ, the spacing is usually 400 mm. If covered with plasterboard, this may be 400 mm for 9 mm plasterboard and up to 600 mm for 12 mm board. For 3 mm hardboard and 12 mm insulation board or wall board, spacing should be 400 mm. For 6 mm hardboard

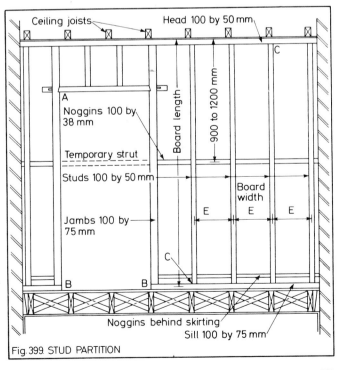

Fig. 399. STUD PARTITION

PARTITIONS

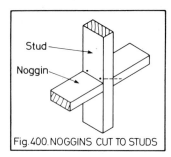

Fig. 400. NOGGINS CUT TO STUDS

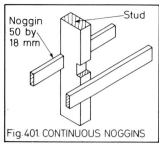

Fig. 401. CONTINUOUS NOGGINS

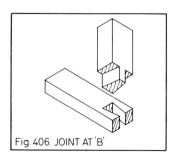

Fig. 406. JOINT AT 'B'

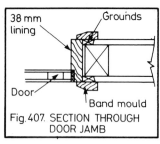

Fig. 407. SECTION THROUGH DOOR JAMB

and 19 mm insulation board, it may be increased to 600 mm.

Generally, supports should be provided at all edges of sheet material. Studs should be spaced within the above limits at distances which are factors of the board width.

Vertical studs need stiffening at between 900 and 1200 mm intervals by horizontal timbers, known as noggins, which may be cut between and skew nailed from the other side of the stud as in Fig. 400. Sometimes the noggins are staggered to simplify the nailing, or continuous battens can be cut flush into the edges of the studs, as in Fig. 401. This will weaken the studs if within the middle third of the height. Noggins should also be placed behind the position of the top edges of the skirtings.

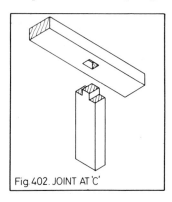

Fig. 402. JOINT AT 'C'

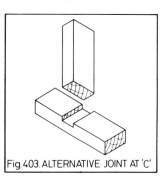

Fig. 403. ALTERNATIVE JOINT AT 'C'

On a cheap job the studs can be cut between the head and sill and toe nailed, but a better job is either the stub tenon Fig. 402 a housing Fig. 403; or the ends of the studs notched over a fillet, nailed to the sill and to the head, as in Fig. 404.

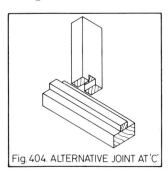

Fig. 404. ALTERNATIVE JOINT AT 'C'

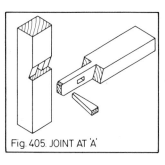

Fig. 405. JOINT AT 'A'

When a doorway is to be fitted, it is recommended that the vertical studs be 75 mm thick to give extra stiffness against the door slamming, Fig. 399. The head over the door is also usually made 75 mm thick but, as it is short, it does not need to be. It is generally secured to the jambs by means of a wedged tenon thrusting against a splayed shoulder which pulls it down to a positive fixing. A similar positive fixing is provided at point 'B' on Fig. 399 where

the jamb fits to the sill. A dovetail joint is best used here (see Fig. 406).

Fig. 407 is a section through a jamb showing the door lining, part of the door and band mould or architrave. Fixing grounds are also shown. These are only necessary if the work is plastered in situ enabling door linings to be fixed after plastering.

TEMPORARY PARTITIONS

It is sometimes necessary to fix temporary partitions where extensions or alterations to an occupied building have to be isolated from the area still in use. They provide security, provide a barrier against dirt and dust, and insulation against sound.

It is easier to make the partition smaller than its eventual accommodation so that it can be fixed with the minimum of damage to ceiling, walls and floor. This is most easily accomplished by wedging down from the ceiling and away from one wall. To avoid damaging the plaster, a board with pieces of felt stuck to it is placed against the wall or ceiling and the partition wedged off it. Cover boards may be used to seal off the joints against the dust.

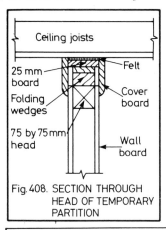

Fig. 408. SECTION THROUGH HEAD OF TEMPORARY PARTITION

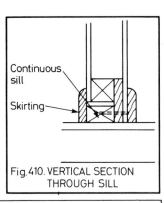

Fig. 410. VERTICAL SECTION THROUGH SILL

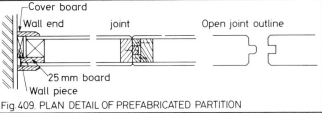

Fig. 409. PLAN DETAIL OF PREFABRICATED PARTITION

Where a number of partitions of standard size are required, it may be worthwhile to manufacture these at the works, in unit panels and fix them on site. Fig. 409 is a part horizontal section and Fig. 410 a part vertical section showing how this may be done with a minimum of machining.

Units are made up of single sheet widths and connected by tongued joints formed with '2' fillets nailed to one end stud and one to the other, the sheeting extending over the joint.

A rebate is formed against the walls, ceiling and floors with a 25 mm board nailed to a reduced stud and this is nailed to a vertical fillet nailed to the wall, ceiling or floor. Cover boards conceal the joints at wall and ceiling positions while the sill is finished with skirtings scribed to the floor.

Partition coverings

Most of the lining or sheeting used to cover partitions consists of some type of vegetable fibre, having the common characteristic that it is hygroscopic and swells and shrinks with changes in moisture content. If expansion takes place after fixing, buckling will be inevitable and to avoid this the moisture content must be increased; i.e. it should be conditioned before fixing. Manufacturers' recommendations are as follows:

HARDBOARD: Apply one half to one litre of clean water to the mesh side of each 1.2 m by 2.4 m sheet of hardboard and brush or sponge the water in.

Leave the dampened boards in the place where they are to be used for 48 hours, stacked back to back on a flat surface. Then fix immediately. If subsequent drying out takes place, they should be re-moistened.

INSULATION BOARD AND WALL BOARD: As these have much more open textures than hardboard, they will take up sufficient moisture if unwrapped and exposed to the atmosphere on site 24 to 48 hours before fixing.

Boards should be stacked open with stickers in the way that timber is stacked for seasoning, i.e. starting with a flat surface the stickers should all be parallel and of uniform thickness and placed exactly one above the other.

Both hardboard and insulation board should be cut from the face side using a fine saw and should be supported close to the cut line.

Fixing all wall boards

It is advisable to allow a gap of at least 3 mm between boards and the spacing of studs and noggins must allow for this, especially on a long run. The order of nailing must be consecutively from the middle down and outwards, as shown in Fig. 411, in numerical order. Nails should be at 100 mm centres around the edges and 200 mm elsewhere. Rustproof nails should always be used and should be of the gauge and length recommended by the manufacturers.

When they are to be driven below the surface and stopped-in panel pins or gimp pins are most suitable. When the joint is covered, lath nails having a greater holding power are suitable. Sometimes nails may have large decorated heads and form a feature of the design. In this case, they must be more accurately spaced, using a long thin batten with saw cuts to suit the shank and spacing required. To get a more accurate finish, holes should be centre punched and drilled in the batten and then cut into from the side.

If the joints are disguised with paper or linen strips, the studs may be arranged with the maximum economy, taking full sheet widths from one end and finishing with a cut board. If, however, the joints are featured, sheets may

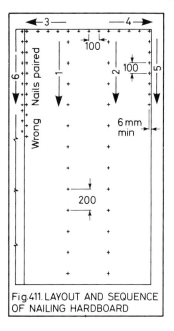

Fig. 411. LAYOUT AND SEQUENCE OF NAILING HARDBOARD

have to be cut to waste to give equal spacing and the studs arranged accordingly.

Fig. 412, 'A' to 'D', gives four ornamental finishes to the joints. 'A' gives a double rib effect, special tools having thin edged sharp cutters acting with a shearing effect are available for this purpose. 'B' gives a double bead effect with wooden beads. 'C' is a vee-joint and is designed to give a clearance while concealing the stud behind. 'D' shows a cover mould which can be moulded to any pattern. Intersections must be mitred-in to reach the full thickness.

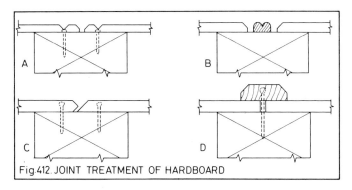

Fig. 412. JOINT TREATMENT OF HARDBOARD

INSULATING PARTITIONS FOR SOUND

Sound insulation in brick and block work is generally achieved by introducing mass into the construction. It can also be obtained as is usual with timber framed buildings separated by internal framing faced with plasterboard and forming a cavity in which is suspended mineral or glass fibre insulating material. This applies in particular to separating walls which also have to be fire resisting.

However, it is often an advantage to have internal walls and partitions with some degree of sound insulations. For example, for bathrooms and toilets. In this simpler construction, a fair degree of sound insulation can be obtained.

1. By isolating the partition from hard contact with the main structure;
2. By isolating the outer skins of the partition from each other;

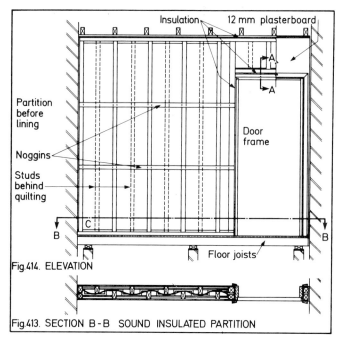

Fig.414. ELEVATION

Fig.413. SECTION B-B SOUND INSULATED PARTITION

3. By introducing one or more intermediate layers of soft material to absorb the sound and reduce the resonance; and

4. By making sure that there are no gaps under the door, or even through the keyhole, for the uninterrupted passage of airborne sounds.

Figs. 413 to 417 show details of a sound-resisting partition having these qualities. There are separate studs to inner and outer skins with a winding space of 50 mm into

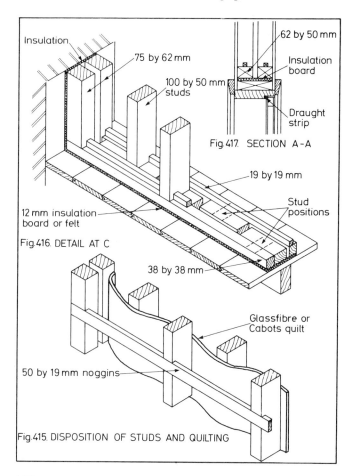

Fig.416. DETAIL AT C

Fig.417. SECTION A-A

Fig.415. DISPOSITION OF STUDS AND QUILTING

which some glass fibre mat or cabots quilting has been introduced (Fig. 415).

The studs are stiffened with continuous noggins separately fixed to the two sides. Each side has separate 38 mm by 38 mm sills widened out with short members as required to take the 100 mm studs. Wall studs are narrower so as not to touch each other. A seating of thick felt or insulation board under the sills, over the heads and behind the jambs, isolate the partition from the main structure. Fig. 416 shows these methods and also the method of fixing studs by notching over a continuous fillet.

The door frame is also isolated from the partition with intermediate layers of insulation board (Figs. 413 and 417) while the short studs over the door are reduced and separated. If the moving joints of the door are stopped with proprietary draught strip, this will cut off the passage of sound and help to deaden any resonance in the door itself.

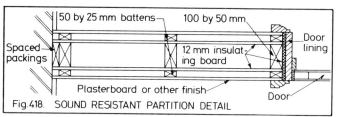

Fig.418. SOUND RESISTANT PARTITION DETAIL

Fig. 418 shows a horizontal section through an alternative type of sound-resistant partition. Insulation board is used as before for isolating the main structure from the building and the door, but the passage of sound through the body of the partition is reduced by introducing two intermediate layers of insulation board. These form direct barriers and also eliminate the linked hard contact between the two outer faces.

The doors in sound-resisting partitions should have the joint to the floor on their bottom edges closed with a rubber strip ploughed in and making firm contact with a threshold when closed.

SCREENS

Screens are, in effect, partitions formed of panelled frames, but which give complete cover, partial cover, or no privacy according to their height. A dwarf screen is usually about 1 m high. It may be placed, say, around an orchestra pit or across an office to give physical but not visual separation between public and private parts. It is made up into long, low panelled frames with a short frame to make a wicket door and is suitably finished with a capping and bed mould under and with a skirting scribed to the floor.

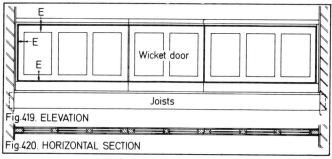

Fig.419. ELEVATION

Fig.420. HORIZONTAL SECTION

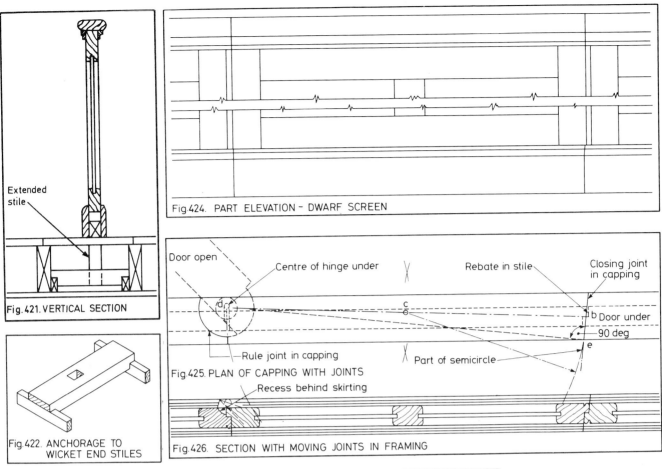

Fig.421. VERTICAL SECTION

Extended stile

Fig.424. PART ELEVATION – DWARF SCREEN

Door open — Centre of hinge under — Rebate in stile — Closing joint in capping

Rule joint in capping

Fig.425. PLAN OF CAPPING WITH JOINTS

b Door under
90 deg
e

Recess behind skirting

Part of semicircle

Fig.426. SECTION WITH MOVING JOINTS IN FRAMING

Fig.422. ANCHORAGE TO WICKET END STILES

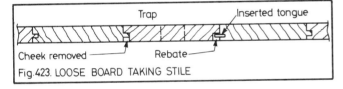

Trap   Inserted tongue

Cheek removed ——   Rebate ——

Fig.423. LOOSE BOARD TAKING STILE

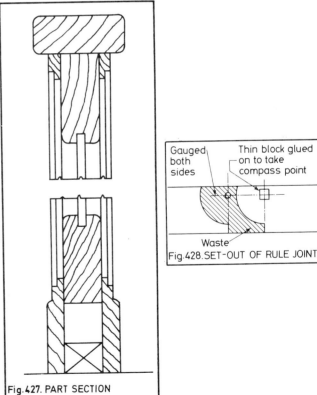

Fig.427. PART SECTION

Gauged both sides   Thin block glued on to take compass point

Waste

Fig.428. SET-OUT OF RULE JOINT

These details are shown in Figs. 419 to 421. The capping serves two purposes:

1. To give a decorative finish;
2. To give lateral stiffness between supports.

If the height is not supported by, say, return panelling or fixing to the end of a counter, it should be held at suitable intervals by some other means. The most common method is with a steel right-angled bracket screwed to the floor and to the panelling. The design may be so arranged that the bracket is covered by a pilaster.

An alternative method, shown here, is to take the end stiles through the floor and stub tenon them to special bearers cut between, and cleated to, the joists as seen in Figs. 421 and 422. A trap should be left in the floor for this purpose. Fig. 423 shows how it may be dust-proofed by cutting away the cheek of the goove on one side and forming a rebate over an inserted tongue on the other. Figs. 424 to 428 show sections and construction in greater detail.

Finishings for screens

In fixed panelling of this type, all the finishings, such as cover moulds and skirtings to be scribed to wall and floor, should be fitted so that the revealed width of all framing (stiles and rails) should be to the standard exposed width (Fig. 419). A feature of this screen is that the skirting, capping and bed mould under the capping all continue

101

across the face of the door. Assuming that standard hinges are being used, this involves some problem in clearance when opening the door.

Although, strictly speaking, the following operations should be carried out in the shop, there is always the chance that capping and skirting may be sent out loose in long lengths. The full details of the joints should be set out on a board and it is essential that the plan centre of the hinge should be accurately placed. The job must be set out and worked to be right first time. Attempts to adjust afterwards to fit will prove extremely difficult.

The opening joint in the capping should be splayed and the splay should be so arranged that it covers the rebated closing joint on the door. A line drawn from 'a' to 'b' (Fig. 425) is bisected and part of a semi-circle drawn about centre 'c' as shown. Then a line drawn from intersecting point 'e' through 'b-e' will give the cut line at right angles to 'a-e' and will thus ensure an opening clearance.

To obtain the set out of the rule joint, draw the small tangent circle at 'a' and draw the open position of the door capping; 'a-p' will now give the radius of the rule joint, to enable the door to open to the angle required. A stop should be fixed early to the floor to limit the door swing and avoid damge to the door capping.

The rule joint should be set out twice on the capping with waste between (Fig. 407). The same gauge and the same setting of the compass should be used and marked both sides of the capping. Joiners' compasses should be used to give a firm scratch mark. The cuts should be sawn roughly to size and then pared in from the scratch mark both sides with paring chisels and gouges. Fig. 426 shows how the closing joint is formed in the framing, how the skirting is cut and the stile recess for the skirting to turn under on the hinging joint.

Higher screens

Figs 429 to 434 deal with a panelled screen taken to about door height. It is assumed that the whole unit can be taken in one piece into the required position (this should be checked at the time of measuring up). The general design and rail positions coincide with other joinery in the vicinity. No skirtings are fitted so there is no hinging problem. To simplify the fitting and fixing, the bottom rails (Fig. 435) are ploughed and drop over fillets screwed to the floor.

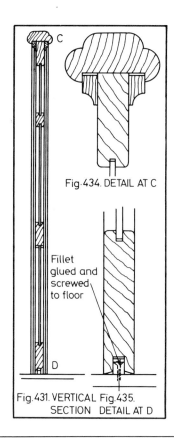

Fig. 434. DETAIL AT C

Fillet glued and screwed to floor

Fig. 431. VERTICAL SECTION   Fig. 435. DETAIL AT D

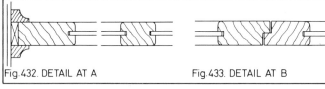

Fig. 432. DETAIL AT A   Fig. 433. DETAIL AT B

This enables the screen to be erected, levelled up so that all horizontal rails are in perfect alignment and scribed both sides to the floor. The groove in the bottom rail should be deep enough to allow this. Lateral rigidity is given by the large capping.

If the top rail and bed mould are too tight to fit the capping groove, the backs of the bed moulds may be

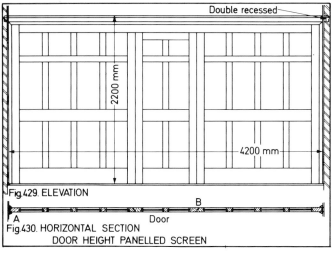

Double recessed

2200 mm

4200 mm

Fig. 429. ELEVATION

B

A   Door

Fig. 430. HORIZONTAL SECTION
DOOR HEIGHT PANELLED SCREEN

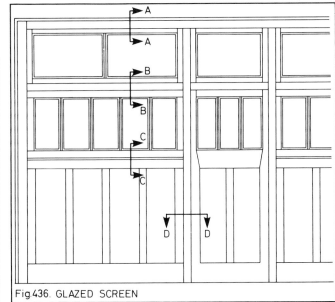

Fig. 436. GLAZED SCREEN

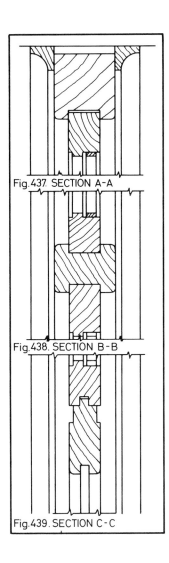

Fig.437 SECTION A-A

Fig.438. SECTION B-B

Fig.439. SECTION C-C

planed to fit without interfering, say, with a polished finish. If the capping is extended from wall to wall and is taken into the wall both sides, the recess one side should be of double depth to allow the capping to slide forward for clearance and then back.

The full height glazed screen is a screen which is fitted into a public building with high ceilings. The glass up to the top of door height is opaque to give light with privacy and above that it is clear. It is assumed that limited access to the area requires that the screen shall be made in separate units and assembled in situ.

The individual frames and the door are glued up in the shop and cleaned up and fitted into a heavy framework which is put together dry in the shop, cleaned off and the parts numbered. It is then taken apart, delivered to the site, and finally reassembled in situ. Figs. 437 to 440 show typical sections. The frames are fitted into grooves in the structural framework.

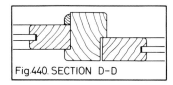

Fig.440. SECTION D-D

Although when carried out properly this is the best job, it means that very careful work is needed in cleaning off the frames and fitting them to the grooves. Then the frames have to be put together with the structural frame and erect in one. In Fig. 440, the side frame is beaded in. This calls for nail holes to be stopped in but enables the frames to be fitted in afterwards without any problems. In the elevation in Fig. 436, detailed in section in Fig. 440, the large sub-frame has been made in two parts. It is presumed that this was necessary due to difficulty of access.

CHAPTER 9

Windows and Doors

WINDOWS

Windows set in the wall in the normal way do not offer many problems and most carpenters are familiar with them. They are generally, of whatever type, weatherproof and efficient within their own construction. Where trouble comes, it is most often due to faults in preparation for and fixing on site.

Parts which are built into the wall should be well painted, particularly on end grain. Where cavity wall construction exists, special care is needed to see that the cavities are well sealed against the jambs and drained clear of the jambs over the lintels. The junction between jamb and brickwork should be waterproofed with a suitable mastic in strip or bulk form or applied with a gun.

Window boards should always be tongued into the sills. They have a tendency to become hollow on the top exposed surface due to unequal drying. To minimise this they should be well painted on the under surface and securely fixed; expecially where there is a free corner, such as with a bay window or combination frame.

It is better to nail them to dovetail fillets in a thick layer of concrete rather than fix to the top row of bricks which is easily disturbed. The heart side of the board should be exposed, thus giving the top surface a tendency to become rounded and balancing the hollowing effect of surface shrinkage. This choice is, however, in the hands of the machinist who prepares the tongue. The cavity under the sill should be sealed with slates or by some other method and the sill should be well bedded with the mortar groove filled.

Cased windows

Cased or double hung windows, although not very popular, do have some advantages. They allow controlled ventilation at different levels and cause no obstruction when open. They are not vulnerable to damage in high winds and also draughts tend to be reduced in these conditions due to the sashes being pressed against the parting and guard beads.

The weights of the sashes have to be compensated, either by coil or spiral spring balances, or by weights on cords over pulleys so that they may be left open in any position. The latter is the traditional method. It is customary to make the weights 0.5 kg heavier than the top sash and 0.5 kg lighter than the bottom one, keeping them up and down respectively in the closed position.

The cords are threaded by means of a strip of lead termed a 'mouse' attached by a fine string to the end of the cord. The mouse, dropped over the pulley and lowered by the string into the casing is pulled out through the respective sash pockets, followed by the cord.

The usual procedure for threading, starting with the hank or cord, is illustrated in Fig. 441 and may be explained as follows:

With the aid of the mouse and string, drop the cord over inside left pulley (1) and out through pocket (2); over inside right pulley and out through pocket (3); over outside left pulley and out through pocket (4) over outside right pulley and out through pocket (5).

The weight is then attached to the cord end, pulled into the casing and up to the outside right pulley and the cord temporarily nailed and cut to length. The next weight, then attached to the free end, is ready to be pulled up to the outside left pulley. The procedure is repeated for the inside cords and weights.

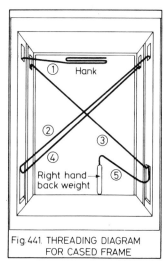

Fig. 441 THREADING DIAGRAM
FOR CASED FRAME

Sash cords will have to be replaced periodically due to breakage. When one cord breaks it is best to renew all the cords while the beads and sashes are removed as they will almost certainly be near the end of their lives.

Bay windows

Bay windows come in a variety of plan shapes but this description is confined to those with splayed ends, which are commonly termed 'Cant Bays'. The same principles of construction apply to other shapes.

Fig. 442 is a part outline plan section above the sill. The ends of the sills are bedded into the side walls and the window boards into the plaster nosings being returned at the ends. Fig. 443 is the part elevation of the sill and Fig. 444 the finish at the head with the cornice, moulding, flat roof and wood rolls prepared for the lead covering. Fig. 445 is a section through the roof over the bay. Bearers to the roof are notched into the BSB (British Standard Beam) and are made up at the top with tapered firrings to give the required roof fall with parallel firrings to the level of the ceiling.

The ends of the bearers are cut to a splay to receive a moulded cornice. To conform to the Building Regulations, the ceiling of the bay only needs to be 2 m high and can be 300 mm lower than the main ceiling.

Fig. 446 shows a section through the head of a bay window where additional height is needed above the window head, using a separate plate supported by short studs. This leaves the window head exposed so a weather moulding becomes necessary. Details of a pitched roof with tiles is shown.

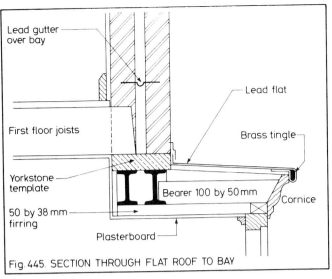

Fig. 445. SECTION THROUGH FLAT ROOF TO BAY

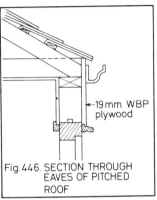

Fig. 446. SECTION THROUGH EAVES OF PITCHED ROOF

Roofing a bay window

To keep all the eaves level, hipped construction then becomes necessary and the carpenter is confronted with the geometry of a miniature but special hipped roof. This is illustrated in Fig. 447.

To set out the roof, draw the plan (A) with the hip mitring the wall plates and the transverse section (B), giving the lengths of the front rafters and the rise with plumb and seat cuts 'a' and 'b'. By projecting from these drawings, the elevation (C) can be drawn. This gives the lengths and plumb and seat cuts 'e' and 'f' of the wall pieces.

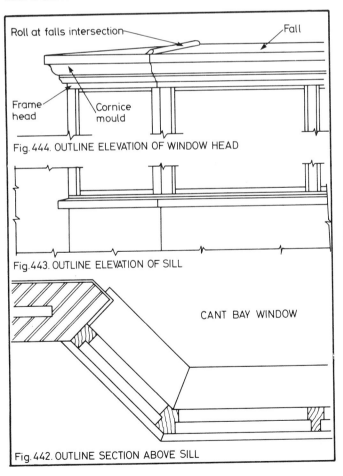

Fig. 444. OUTLINE ELEVATION OF WINDOW HEAD

Fig. 443. OUTLINE ELEVATION OF SILL

CANT BAY WINDOW

Fig. 442. OUTLINE SECTION ABOVE SILL

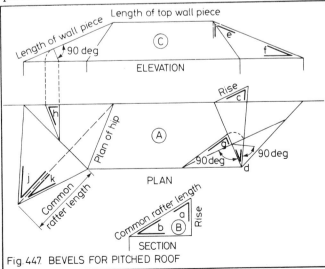

Fig. 447. BEVELS FOR PITCHED ROOF

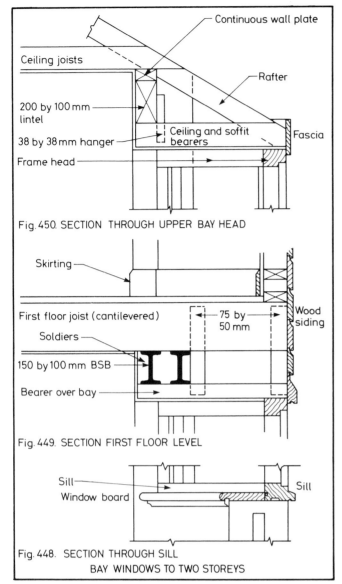

Fig. 450. SECTION THROUGH UPPER BAY HEAD

- Continuous wall plate
- Ceiling joists
- Rafter
- 200 by 100 mm lintel
- Ceiling and soffit bearers
- 38 by 38 mm hanger
- Fascia
- Frame head

Skirting

First floor joist (cantilevered)
75 by 50 mm
Wood siding
Soldiers
150 by 100 mm BSB
Bearer over bay

Fig. 449. SECTION FIRST FLOOR LEVEL

Sill
Window board
Sill

Fig. 448. SECTION THROUGH SILL
BAY WINDOWS TO TWO STOREYS

The edge bevel of the wall piece, to lie in the roof plane, is shown in 'h' and is self-explanatory. With reference to (C) and (A), the rise square to the hip plan in (C) gives the hip plumb and seat cut 'c' and 'd', and the dihedral angle square to this with its horizontal trace cutting the wall plates gives the backing bevel of the hip at 'g'.

If the length of the triangle of the hipped end in A is extended to the common rafter length this will give the jack rafter cuts 'j' against the wall and 'k' against the hips.

When a bay window extends to two storeys or roof height, the problem arises of supporting the first floor joists which must not be carried by the window head and the finish of the window at roof level.

Fig. 448 is a section through the window sill which is typical of most bay window construction. Fig. 449 is a section through the bay window head to ground floor. The BSBs over the bay in the main wall position carry the floor joists which cantilever over the ceiling suspended from them.

Vertical studding extends to the first floor window with rebated shiplap boarding nailed to it on the outside and plasterboard on the inside. A moisture barrier of, say, building paper is placed under the shiplap on the outside and a vapour barrier on the inside under the plasterboard.

Intermediate insulation within the cavity is also necessary. Note the protection provided to the window head at ground floor level.

Where the head of the bay to the first floor extends to roof level, this could require quite a complex finish if the bay extended beyond the line of the main roof. It could involve construction on the outline of Fig. 447, or the extension beyond the eaves could be continued as a flat.

The simplest solution is to bring the fascia of the roof in line with the outer head of the bay as shown in Fig. 450. This would need some adjustment in the position of the wall plates if, say, a hipped end had a different overhang.

Skylights

Skylights are usually provided in attics and lie parallel to the roof plane. With some loss of light they could be boxed down to a laylight in the ceiling of the upper floor.

The opening to the roof is formed with trimmed rafters supported by trimmers in turn carried by trimming rafters, as in Fig. 451. The bending stress on the latter depends upon the position of the opening and is greatest when this is at the centre of the span between purlins. The joint at 'A', shown in Fig. 452, has the advantage that it can be pulled up tight with the wedge. Fig. 453 shows other methods which are in general use.

Trimmers and trimming joists are best made 25 mm thicker than common rafters. The shaded areas in 'a', 'b' and 'c' in Fig. 454 show bending stress distribution in a rectangular timber beam; the black areas show the relative loss of strength through cutting away to form joints as similarly lettered in Figs. 452 and 453.

Fig. 455 gives a longitudinal section through the skylight. The sash stands on a curb which lifts it clear of the roof and above the flow of water. The lower member of the curb is vertical to give the maximum of light.

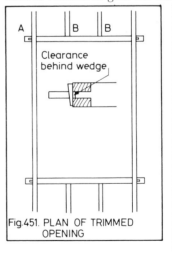

Fig. 451. PLAN OF TRIMMED OPENING

Clearance behind wedge

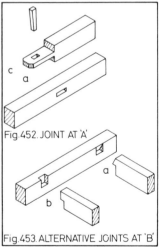

Fig. 452. JOINT AT 'A'

Fig. 453. ALTERNATIVE JOINTS AT 'B'

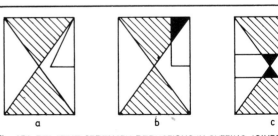

Fig. 454. RELATIVE STRENGTH REDUCTIONS IN CUTTING JOINTS

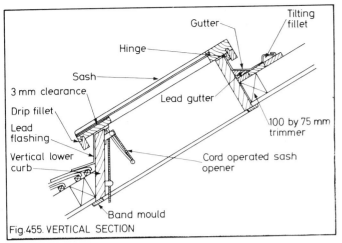

Fig.455. VERTICAL SECTION

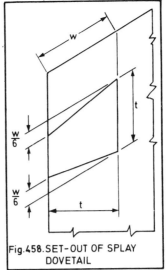

Fig.458. SET-OUT OF SPLAY DOVETAIL

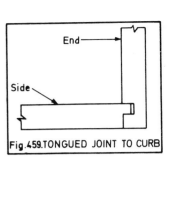

Fig.459. TONGUED JOINT TO CURB

Construction details of the sash are mainly the concern of the joiners' shop and are not dealt with here. The top of the sash is shown hinged. The hinges and screws should be of brass. A gutter is placed at the back of the skylight with lead flashing over the tiles at the front.

Fig. 456 is a cross-section normal to the roof slope with alternative methods of forming a drip. A side gutter which takes the water away from the roof behind the skylight is also shown.

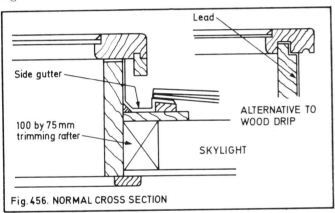

Fig. 456. NORMAL CROSS SECTION

Fig. 457 shows the method of joining the corners of the curb with dovetails. For maximum security of joint, it is better to have a pin inside each edge than the end of a dovetail as there are two keyed and glued surfaces instead of one and the joint will be less likely to loosen through uneven shrinkage. The dovetail splays must always be related to the grain of the timber and not necessarily to the jointed end. This is indicated in Fig. 458 which shows the method of setting out one pin. Fig. 459 is an alternative method of jointing the curb using a tongued and grooved joint.

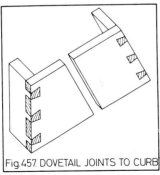

Fig.457. DOVETAIL JOINTS TO CURB

When the skylight has to be set in a flat roof, the slope necessary to give the positive drainage of the sash requires that the sides of the curb shall be steep triangles and the back very wide. Solid construction in this case is not a practical proposition and an alternative is the use of lined framing. Fig. 460 is a section through a skylight in a flat roof taken parallel to the slope. Side and end frames are made of 38 mm timber tenoned together. It is covered on the inside with 9 mm WBP (weather and boil proof) plywood, and on the outside with vertical matchboarding.

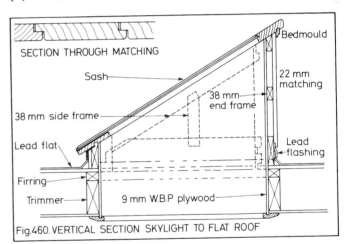

Fig.460. VERTICAL SECTION SKYLIGHT TO FLAT ROOF

Rebated joints in the matchboarding are better than tongues and grooves as the parts of the joints are more robust and less inclined to retain water or break under exposure to weather extremes. The flashing over the turned-up lead of the flat is tucked under the matchboarding which is rebated to receive it. The joint between the curb and the sash are closed with a tongued bed mould. A drip is not necessary at the top but should be formed at the sides. The plywood lining extends below the curb to the ceiling level, the edge being covered with a grooved moulding.

Dormer windows

Dormer windows are vertical lights in a pitched roof necessary to give light and air to a room when it is fully or partly in a roof. They are generally an essential part of

107

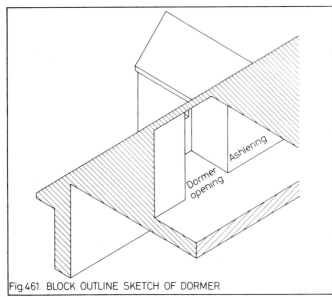

Fig.461. BLOCK OUTLINE SKETCH OF DORMER

a loft conversion plan. Fig. 461 is a block outline sketch from the inside of a dormer window in a loft. Vertical walls, cutting off the lower part of the loft from the room, are formed with vertical studding known as ashlaring. The head of the studding forms a plate over which the common rafters are notched. As the window within the dormer framing will have its sill from 750 mm to 1000 mm from the floor, this will have to be set back towards the eaves and an opening formed as shown.

The roof of the dormer extending out from the main roof may be of any type but is usually either flat or pitched with a gable or hipped end to agree with that of the main structure. In Fig. 461 a gable end is shown. Fig. 462 shows section 'A-A' of the front elevation of a dormer window (Fig. 463) but without roof covering. Fig. 464 is a pictorial sketch showing the general arrangement of the roofing and dormer timbers. These three drawings should be studied together.

The framing to the dormer rests on sills over the flooring laid over the required area. The framing is housed into the

head and sill and the trimming rafters notched to receive it. The trimming rafters are made out to the inside studding with packings. The trimmed rafters at the foot are cut against the front vertical studs.

Rafters to the dormer head are birdsmouthed over it and nailed to the ceiling joists. An additional short rafter to the main roof may be necessary to take the ridge. The roofing battens and jack rafters are cut on to and fixed to a layer board nailed to the main rafters at the correct angle.

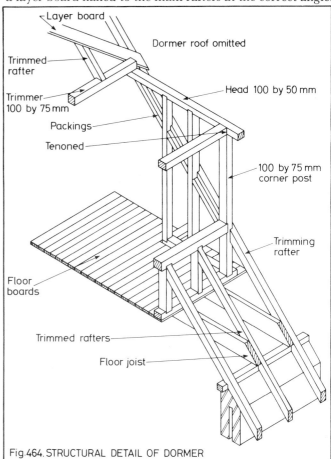

Fig.464. STRUCTURAL DETAIL OF DORMER

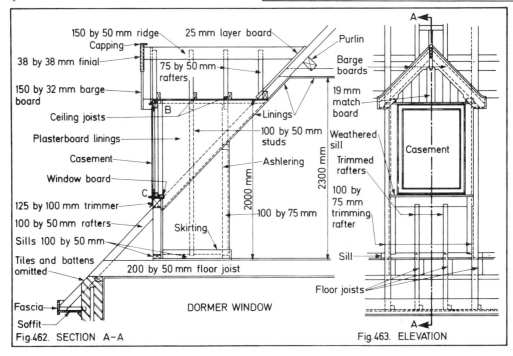

Fig.462. SECTION A-A

DORMER WINDOW

Fig.463. ELEVATION

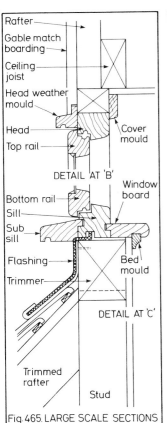
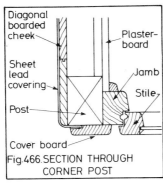
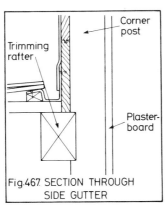

Fig.466.SECTION THROUGH CORNER POST

Fig.465.LARGE SCALE SECTIONS

Fig.467. SECTION THROUGH SIDE GUTTER

The end of the gable is finished with a barge board tongued to a finial which is also recessed to receive the ridge. A capping to the barge board covers the tile ends. The upper trimmed rafters are birdsmouthed over a trimmer, itself supported by the ends of the heads and the gable end is matchboarded.

Details of finishings are given in Figs. 465 to 467. In the detail at 'B' on Fig. 465, (see also 'B' Fig. 462), the head joint is covered by a weather mould over which the matchboarding is rebated. In detail at 'C', flashing is tucked into the grooved sill and taken down over the tiles. The cheeks of the dormer are covered in lead taken round to the front under a rebated cover board (Fig. 466).

The bottom edge is extended over a side gutter and lead lined. Sheet lead may be secured to the timber cheeks in

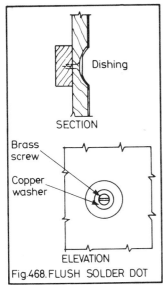

Fig.468.FLUSH SOLDER DOT

various ways. One method, in which the carpenter may be involved, is by means of flush solder dots. A hollow is dished out in the boarding and supported at the back by a block. The lead is dressed into it, secured with a screw, and the hollow filled flush with solder.

Lantern windows

Figs. 469 and 470 are the vertical section and plan of a lantern light in block outline. They are more commonly confined to flat roofs and bear some resemblance to a low greenhouse with hipped roof lights and side lights. Fig. 471 shows selected plan details of the trimmed opening in a roof to take a lantern (Fig. 472 is the vertical section).

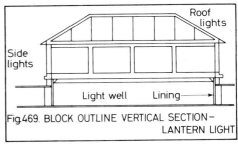

Fig.469. BLOCK OUTLINE VERTICAL SECTION– LANTERN LIGHT

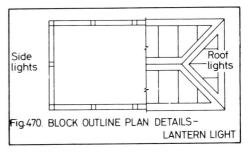

Fig.470. BLOCK OUTLINE PLAN DETAILS– LANTERN LIGHT

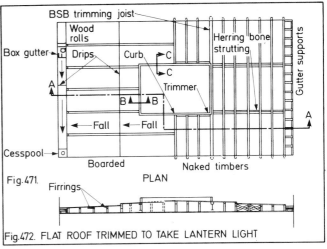

Fig.471. PLAN

Fig.472. FLAT ROOF TRIMMED TO TAKE LANTERN LIGHT

It is assumed that the weight of the lantern is carried by two steel beams with wood trimmers cut between them. Starting at the gutter ends with 200 mm by 50 mm joists, they are built up with firrings to give the necessary falls and drips to a depth of 300 mm at the centre.

The sills of the lantern lights are carried on curbs which lift them about 125 mm off the roof. In Fig. 473, (section 'BB' Fig. 471) the curb is taken down to the top of the BSB and secured, as shown, by packings and hangers.

The inside of the light well is lined with panelled framing down to the ceiling level. Condensation, collected on the underside of the roof lights, runs out mostly under the

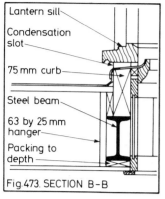

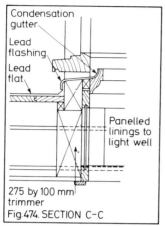

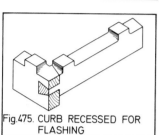

Fig.473. SECTION B-B

Fig.474. SECTION C-C

Fig.475. CURB RECESSED FOR FLASHING

continuous presence of water between storms, due to it being trapped or confined between two enclosed surfaces. Anti-capillary grooves should, ideally, be as near to the outside surface as possible but joints on either side should not be so open as to permit rain being blown through by the wind.

Some steps should be taken to ensure that water running down the door surface cannot be collected or run back at threshold level. Figs. 476 and 477 show some common precautions to avoid water penetration. In Fig. 476, the weather-board extends over the step so that the water over the weathering is checked by the drip clear of the step. In Fig. 477, a water-bar in the stone threshold with a mat well behind will check the water blown up the weathered surface while it is prevented from creeping into the rebate by a groove in the jamb. Fig. 477 together with Fig. 478 show how the door stop is shortened to allow the collected water to run away.

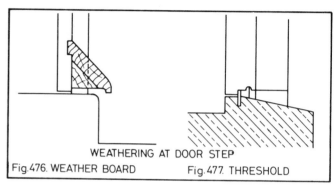

WEATHERING AT DOOR STEP

Fig.476. WEATHER BOARD   Fig.477. THRESHOLD

glass over the apron rail in a space provided. However condensation collected by the side lights has to be disposed of. It is usually collected in a gutter below the sill, which is moulded to appear as a cornice. This is drained out on to the roof, either through copper tubes taken through the curb, or through lead lined shallow sinkings formed at intervals in the curb.

Fig. 474 shows section 'C-C' of Fig. 471 through the wood trimmer between steel beams. Otherwise construction details are similar. Fig. 475 shows the joint between the curbs, members and the sinkings for drainage of condensation.

EXTERNAL DOORS

External doors may be of a wide variety of designs such as framed; ledged, braced and battened; panelled; wholly or partly glazed; or flush. However, in most cases, the frames are similar, of robust construction of 100 by 75 mm section or larger.

They may be fixed in a prepared opening in the wall; in which case they are nailed to plugs or pads previously built into the horizontal brick joints. Alternatively they may be erected and plumbed and the brickwork built around them, anchorage being obtained at not more than 600 mm vertical intervals from wrought iron right-angled straps screwed to the jambs and built into the joints.

The bottoms of the jambs should, preferably, be secured with iron dowels in the sill, or a shoe of the same section should be fitted to the end of each jamb. A round or square dowel on the end is set into the sill.

The joints between the wooden framework and brickwork should be sealed on the weather side with a suitable mastic. The bottoms of the jambs should be well painted. If the end grain is first saturated with an organic solvent preservative, such as Cuprinol, which is allowed to dry before painting, extreme durability will be ensured. It is important that all timber surfaces in contact with the brickwork should, at least, be well painted.

One of the main causes of decay in woodwork is the

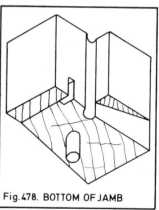

Fig.478. BOTTOM OF JAMB

It is important, in relation to the fixing of door frames—whichever method of erection is used—to ensure that, initially at least, there is a gap of 6 mm to 12 mm between the door head and the lintel. Otherwise the inevitable subsequent settlement of the brick joints will throw the weight of the upper wall and lintel on to the door jambs, probably also causing cracks in the brickwork itself.

In the fixing of door frames, precision is all-important if subsequent door hanging problems are to be avoided. It is sound policy to check door jambs with a long straight-edge before erecting; a pronounced bow can usually be corrected with an additional brace.

INTERNAL DOORS

The treatment of internal door frames and linings depends upon the partition wall thickness and constitution. In

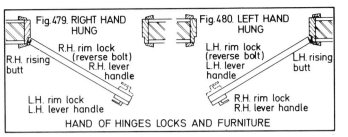

Fig. 479. RIGHT HAND HUNG

R.H. rim lock (reverse bolt)
R.H. lever handle

R.H. rising butt

L.H. rim lock
L.H. lever handle

Fig. 480. LEFT HAND HUNG

L.H. rim lock (reverse bolt)
L.H. lever handle

L.H. rising butt

R.H. rim lock
R.H. lever handle

HAND OF HINGES LOCKS AND FURNITURE

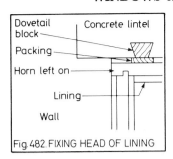

Dovetail block
Concrete lintel
Packing
Horn left on
Lining
Wall

Fig. 482. FIXING HEAD OF LINING

112 mm walls, it is usual to provide linings about 38 mm thick with rebates for the door, as in Fig. 479; or 22 to 32 mm thick with planted stops.

The width of the linings extends the plaster thickness either side of the wall, the joint being covered by an architrave mould mitred around the opening. On high-class work, rough grounds, straightened and nailed to the wall form a stop to the plaster, the linings being fixed after plastering. Traditionally, when plaster was 19 mm thick, they were nailed to plugs in the wall face. With modern finishing plaster being only 12 mm thick, the width of the opening may be increased to accommodate the grounds within the wall thickness as in Fig. 492.

In common work, the opening is usually from 12 to 25 mm wider than the linings to allow for straightening against stock brickwork. Fixing can be either by nailing to wood pads bedded in the horizontal brick joints, or to plugs driven into joints, which previously were cleaned out with a plugging chisel. Plugs should be chopped with an axe, twisted, but of parallel thickness (not tapered), and a drive fit into the joint.

Plugs and pads should be left projecting into the opening to be cut as necessary. The bottoms of the jambs are usually left long and should be cut off to the required height of the door. It is good policy to check the floor with a spirit level across two parallel blocks and to make one side shorter by the amount that the floor is high.

Before cutting back the surplus on the pads, both jambs of the opening should be checked for regularity and general plumb. It may be an advantage to have more clearance on one side than the other. Then the pads should be marked on one side with a plumb rule and the surplus cut off.

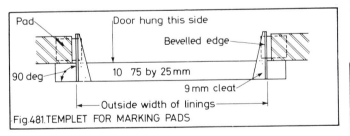

Pad
Door hung this side
Bevelled edge
90 deg
10 75 by 25mm
9 mm cleat
Outside width of linings

Fig. 481. TEMPLET FOR MARKING PADS

The pads should also be cut square to the line of the wall. To ensure this, they can first be marked with a simple templet, as shown in Fig. 481. Once the pads have been cut one side and checked for plumb and straightness, the others can be marked out from the templet and also cut, the distance being the width over the outsides of the linings.

The horns on the head of the lining should be left as long as possible because this will increase the strength of the housed joints and increase the rigidity (see Fig. 482). The heads of the linings can be fixed to embedded wood fillets in the concrete lintel, the head being left slightly hollow

to allow for settlement of the lintel. If the settlement does not take place, they can always be tapped down before plastering.

Accommodating inconsistencies of the fabric

When plumbing the linings across the wall thickness, a little discretion is sometimes needed; this is particularly so in modifications to old buildings, as is popular today, in which walls may be out of plumb and floors out of level. In this case a self-satisfied insistence on precision can leave someone following on with a headache. The only really sensible way to deal with a problem of this nature is to divide the major and outstanding error into as many minor and insignificant ones as possible.

For example, assume that the combined effect of inaccuracy of floor and wall meant that, in order to open to 90 deg., the bottom of the door would have a gap of 25 mm under its closing edge when shut. The following chain of inaccuracies could make the error unnoticeable:

1. The lining could be fixed 3 mm forward at the bottom and 3 mm back at the top, leaving the plasterer some adjusting which he could do over a wide area.
2. Hinges could be fitted with a maximum projection or throw at the bottom and none at the top (see Fig. 484). A tilt could be given to the door when open only, as in Fig. 483.
3. The bottom of the door could have, say, an extra 3 mm clearance off the floor and then shot away a little more on the closing end until it just cleared the carpet.

If the doors and linings are square, edges straight and sizes correct, little work is involved in fitting the door beyond cutting off the horns and shooting the edges clean. Often this is not the case and the writer recommends the following procedure. This anticipates that there will be

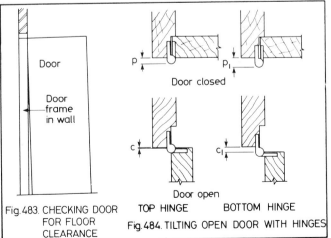

Door
Door frame in wall

P
Door closed
P_1

c
Door open
c_1

TOP HINGE   BOTTOM HINGE

Fig. 483. CHECKING DOOR FOR FLOOR CLEARANCE

Fig. 484. TILTING OPEN DOOR WITH HINGES

inaccuracies in all directions based on the principle of mark and cut rather than continued trial and error, which is time-consuming.

1. Check jambs with a straight edge and note if either is in bad shape.
2. Take the width of the opening at the top and bottom and mark them on the door extremities so that there is an equal margin for planing both sides.
3. Take the height of the opening and check against door height. If the door is short and is framed with horns, they can be left long and a make-up planted between at the bottom.
4. Cut the bottom of the door off square, clear of the bottom rail and stand against the opening tight to the jambs so that the previous marks coincide with them. Mark both sides from top to bottom and along the top with a fine pencil.
5. Shoot the stiles accurately to the pencil lines, keeping the edges dead square, when the door should just fit for width.
6. Check the average width of the stiles and mark this upwards on the top rail. The distance between this and the pencil line will give the amount the door has to drop. Place the door in position again and, taking the above measurement plus whatever is allowed for clearance, say 6 mm, scribe along the floor and saw and plane to the line.
7. Lift the door into position with the correct clearance and mark the top rail, then cut and plane.

If the work has been done carefully, it only needs a few shavings off to give the penny joints and a slight closing bevel. All keen edges should be finally removed.

Fitting the hinges

Fitting of hinges is a straightforward matter in which care saves time. The usual positions are 225 mm from the bottom, 150 mm from the top of the door and the third, if necessary, midway. They should not be placed where screws will meet end grain of tenons and should coincide with hinge blocks on flush doors.

Floor clearance should be checked by standing the door at right-angles to the jamb, as in Fig. 483. Fig. 484 shows how the throw of the hinges can be used to tilt the bottom edge while still keeping the door flush when shut.

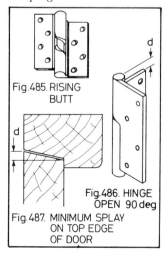

Fig. 485. RISING BUTT

Fig. 486. HINGE OPEN 90 deg

Fig. 487. MINIMUM SPLAY ON TOP EDGE OF DOOR

Rising butts (Fig. 485) can be used to lift the door over the carpet as it opens, provided the edge is some way from the doorway (i.e. not fitted). As the door does not leave the rebate until it has opened nearly 90 deg., the amount of lift over this angle of movement must control the necessary bevel on the top edge (see Figs. 486 and 487). The rules for determining the hand of hinges and other fittings to doors are frequently given a complex interpretation. The writer tries to simplify them as much as possible as follows:

1. Stand on the side of the door where the hinge knuckle shows, then the position of the hinge is the hand of the door. For some reason, however, a left-hand rising butt fits a right-hand door and vice-versa.
2. For locks and other fittings e.g., handed lever furniture on the face of the door, stand on the side of the door on which the fitting is, then the position of the hinge gives the hand of the fitting.

Check with the diagrams.

Doors in partitions

As partitions 75 mm thick or less tend to lack rigidity, openings for doors in them constitute a weakness which must be countered. Door jambs are therefore carried up to full storey height, where they obtain support from the ceiling joists and, in turn, stiffen the partitions.

This is illustrated in Fig. 488. Right-angle wrought iron straps, as previously described, are screwed to the frame and incorporated in the wall as it is carried up. The jambs are wide enough to embrace the plaster finish which

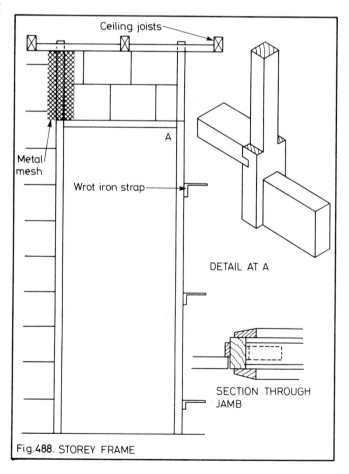

Ceiling joists

Metal mesh

Wrot iron strap

A

DETAIL AT A

SECTION THROUGH JAMB

Fig. 488. STOREY FRAME

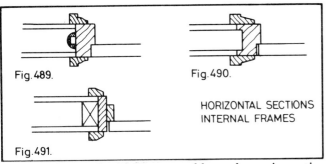

Fig.489.   Fig.490.

HORIZONTAL SECTIONS
INTERNAL FRAMES

Fig.491.

includes the usual architrave mould, as shown in section through jamb and are reduced to nett wall thickness as shown in detail at 'A' in Fig. 488.

Other methods of retaining the wall are also shown. In Fig. 489, a key into the block is formed with a fillet nailed to the jamb. In Fig. 490, the jamb is housed out to receive the walling. In Fig. 491, a separate 'buck frame' is erected to contain the wall and door linings are nailed to it.

Swing doors

Doors which swing feeely between jambs returning automatically to the closed position can be hinged to the jambs with various types of spring hinges. Those moving through 90 deg. are termed 'single' action; and those which move both ways through 180 deg. are termed 'double' action.

As there is no special provision for adjustment, they usually have to be accurately fitted to bring the doors meeting stiles into line. The top hinge only needs to be spring-loaded, the bottom one being blank.

Probably the most common and best method of hanging swing doors is with floor springs contained in a box set flush in the floor. The compression springs act on a cam and incorporate an oil leak positon to steady the return of the door through the last 100 mm or so. A pivot in the head takes the top of the door. This is adjustable for plumbing the door stile. A shoe, receiving the bottom corner of the door, fits over a square spindle operated by the spring. A screw in the side of the shoe gives limited adjustment for aligning the doors.

It is essential to check that there is sufficient depth in the floor to take the box mechanism during the early part of the construction. Fig. 492 shows the set-out necessary to the smooth working of the pivoted jamb. The centre should be set-out so that the door clears the jamb by about 6 mm when the door is open at right-angles. Meanwhile, the recess to take the round hanging edge should be hollowed to a shorter radius to give clearance when moving and a tight joint when closed. It is especially important to ensure that

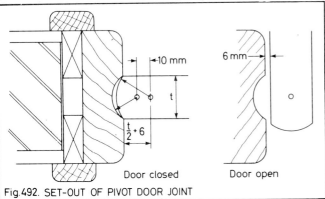

←10 mm   6 mm→

t

$\frac{t}{2}$ + 6

Door closed   Door open

Fig.492. SET-OUT OF PIVOT DOOR JOINT

both door jambs are set plumb so that the meeting stiles come into line and the floor spring and upper pivot are correct in relation to them.

It helps in the correct positioning of the floor spring if a length of timber of the correct length is placed between the shoes when setting the boxes.

Sliding doors

Sliding doors come in three main groups: i.e. straight sliding, round the corner, and folding. In deciding which type to use, the designer is governed by the space and position available for them when opened. Thus, in Fig. 493, one door can slide open to its full width. In Fig. 494, half-doors slide open both ways.

In Fig. 495, only one half-space is needed at one time as for a double garage; while in Fig. 496 limited widths means that the open doors must be in banks of two or more in depth. In Fig. 497, doors are taken round a quadrant to finish along a side wall; while in Fig. 498 and 499, accommodation in depth is sacrificed to give maximum clear open widths between two side walls with no front cover.

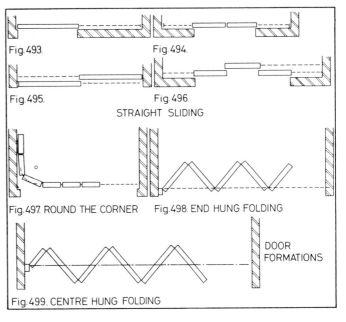

Fig.493.   Fig.494.

Fig.495.   Fig.496.
STRAIGHT SLIDING

Fig.497. ROUND THE CORNER   Fig.498. END HUNG FOLDING

DOOR FORMATIONS

Fig.499. CENTRE HUNG FOLDING

The doors may slide in floor tracks on trolleys set in the bottom of the doors with top guides running in channels, or may be hung from overhead tracks with floor channels to take the bottom guides. Again the channel may be in the bottom of the door and the guide fixed to the floor. Details of these are available in considerable detail in various trade catalogues and will not be discussed here.

Figs. 500 and 501 show plan and elevation of a single domestic sliding door with details to give the finished appearance expected of this type of work.

The overhead track, which may carry rollers or ball sheaves, is fixed to the wallpiece and is completely enclosed by a pelmet or casing (Fig. 502). The brick jambs are cased in linings with architrave mould finish on the back. There are vertical jambs forming the stops for the open and closed positions of the door on the face (Figs. 503 and 504); while a cornice mould to the pelmet is returned to the face of the wall.

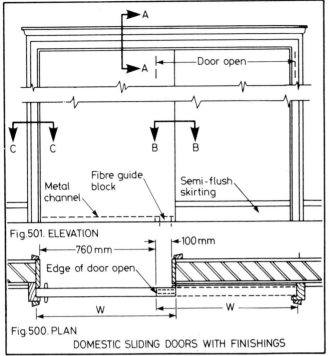

Fig.501. ELEVATION

Fig.500. PLAN

DOMESTIC SLIDING DOORS WITH FINISHINGS

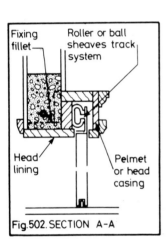

Fig.502. SECTION A-A

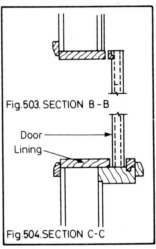

Fig.503. SECTION B-B

Fig.504.SECTION C-C

Note that the door edge in the open position must stand forward about 100 mm to make the handle available on the back, and this width must be added to the opening to make it 860 mm wide if the customary 760 mm open access is required.

The bottom of the door is ploughed to take a metal channel which is a sliding fit over a fibre block screwed to the floor at the inner extremity of the open door. If the door has to be secured, a bird's-beak lock will be required. This has a double closed hook which passes through, and opens behind a slotted plate in the jamb.

Larger sliding doors

Large doors to industrial buildings and garages may be suspended from, or run on, tracks of a wide variety of types and sizes. The details of this equipment will govern the layout of the fixings provided, and for this the manufacturer should be consulted.

If doors are straight sliding, trolleys may be rigidly attached to the top of the door, by means of horizontal plates or double right-angle vertical straps which overlap

each other to embrace the door thickness, and slotted to contain the retaining nut of the trolley.

Height adjustment is always available to enable the doors to be levelled. The channel taking the guides of overhead hung doors may have the bottom rounded for easy cleaning. It may also be of excess depth to maintain sliding clearance when it is only roughly raked out.

When doors have to slide around a quadrant track, as in Fig. 505, the trolley and guide are combined with the top and bottom hinges (Figs. 509 and 510) so that suspension is at the door edges. One or more intermediate hinges must also follow the pattern of the trolley hinge so as to keep the centres in alignment.

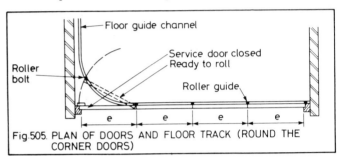

Fig.505. PLAN OF DOORS AND FLOOR TRACK (ROUND THE CORNER DOORS)

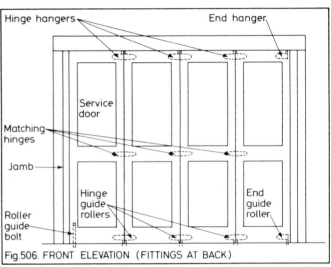

Fig.506. FRONT ELEVATION (FITTINGS AT BACK)

Fig. 508 is an outline vertical section showing one type of tubular track supported by enclosing brackets bolted to wall-pieces fixed to the wall. Note that the detailed dimensions of the track and hanger together with the thickness of the door govern the thickness 't', of the wall-piece in order to give the working clearance 'c', from the lintel. The wallpiece to the return track will have to be much thicker or be packed off the wall to give clearance for the roughness of the brickwork and the projection of any locks or furniture (Fig. 507).

In order to give complete closure, the end door, which is usually a service door, must leave the quadrant and fold back into the rebated jamb where it may be secured with any normal type of door lock. The top of the door is therefore left free (without a hanger) and the bottom is fitted with a guide bolt which drops into the floor channel when the doors are to be slid open.

Figs. 505 and 506 show these details in plan and elevation. Fig. 507 shows the plan of the track and brackets and the method of supporting the centre of the overhead quadrant track.

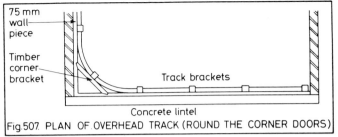

Fig.507. PLAN OF OVERHEAD TRACK (ROUND THE CORNER DOORS)

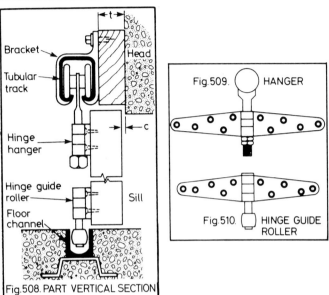

Fig.508. PART VERTICAL SECTION

Fig.509. HANGER

Fig.510. HINGE GUIDE ROLLER

It is essential that the centre of the overhead track is plumb over the centre of the floor guide channel. Fig. 511 shows a stout board templet by means of which this may be achieved. Note that the centre of the floor quadrant should be plumbed normal to the curve.

So far, it has only been necessary to ensure that the overall width of the door or doors should agree with the

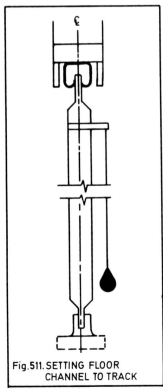

Fig.511. SETTING FLOOR CHANNEL TO TRACK

space into which they have to fit. With centre-hung or end-hung doors (Figs. 498 and 499), it is also essential that, when the doors are open into the folded position, all the hinges and hangers should also be in alignment. Therefore, not only must the overall width be correct, but also the individual widths must be correctly related.

Ideally, the doors should be set out, fitted together and prehinged in the shop so that, it should only be necessary to put them together again on site and perhaps take a few shavings off the closing edge of the last door.

If this has not been done, each individual door will have to be brought to its correct size. To do this by trial and error is likely to bring trouble and it is preferable, and easier, to calculate exact sizes (working to 0.1 mm which is judged in marking).

First, taking the end-hung door (Figs. 498, 512 and 513), sketch the plan of the door folded (Fig. 512). This should include the centre of the jamb hinge, the centre line of track and 'm' which is the distance the centre of the end hanger must come inside the meeting stile to allow clearance on overhead trolleys. Assuming that the doors fold from both sides, measure the distance between the jamb rebates and divide by two (for centre joint).

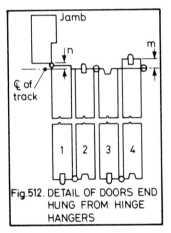

Fig.512. DETAIL OF DOORS END HUNG FROM HINGE HANGERS

Find the exact measurement of 'n' and 'm'. Then reason thus:

If all leaves were the same width and 2 and 3 were correct, overall width would be n + m short. Therefore, deduct this sum from the overall width and divide by 4. This will give the correct (tight) width for leaves 2 and 3. Then add 'n' to leaf 1 and 'm' to leaf 4. As a check, add the lot together.

EXAMPLE: Width between jambs = 7423 mm, measurement 'n' = 6 mm and measurement 'm' = 13 mm.

Then width of opening to centre line

$$= \frac{7423}{2} = 3711.5 \text{ mm}.$$

Then width of leaf 'w' or

$$3 = \frac{3711.5 - (6 + 13)}{4} = 923.1 \text{ mm}.$$

Leaf 1 = 923.1 + 6 = 929.1 mm.
Leaf 4 = 923.1 + 13 = 936.1 mm.

Check overall width = 929.1 + 923.1 + 923.1 + 936.1 = 3711.4 mm.

This shows an overall error of 0.1 mm which is not practically measureable. These are tight dimensions and

115

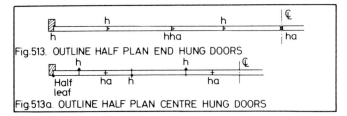

Fig.513. OUTLINE HALF PLAN END HUNG DOORS

Fig.513a. OUTLINE HALF PLAN CENTRE HUNG DOORS

the thickness of the joint between leaves must be decided upon and deducted from each leaf.

In Fig. 514, the leaves are hinged together with normal butts and suspended from hangers set in clear from the end of each leaf. Three leaves are the same and one is short by distance 'd'. Therefore, add 'd' to the half-span taken as before, divide by four, and deduct 'd' from the first leaf; then check as before.

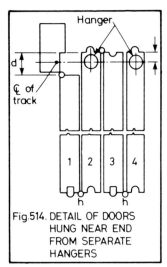

Fig.514. DETAIL OF DOORS HUNG NEAR END FROM SEPARATE HANGERS

Figs. 513 'a' and 515 show centre-hung leaves. The first one is less than half the others by the distance 'd'. Add 'd' to half width of opening as before. Divide by seven. Multiply the result by two for full leaf width. Deduct 'd' from the result to give leaf one. Check by adding all leaf widths together.

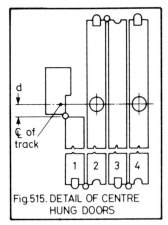

Fig.515. DETAIL OF CENTRE HUNG DOORS

Fig. 516 shows folding doors with rebated joints. In this case, it should be noted that the half partition extends half a rebate, 'e', over the centre line, so this is the width used. When the door is folded back, all the centres of the hinges must be in line. So that all the full leaves may be the same width, each leaf must be the width between hinges, plus one rebate.

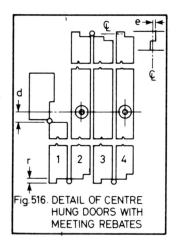

Fig.516. DETAIL OF CENTRE HUNG DOORS WITH MEETING REBATES

The working-out procedure is, therefore, as follows:

1. Add distance 'd' + 'e' to the half width of doorway.
2. Divide by seven to give half width between hinges.
3. Deduct 'd' and add 'r' for width of first small leaf.
4. Double half-width and add 'r' for all full leaves.
5. As a check, deduct 'r' from each leaf before adding together as this does not increase the width.

When folding doors and partitions are used in public rooms, etc., the overhead track will have to be cased as in Fig. 517. One side of the casing should be screwed on so that it can be removed to get at the track.

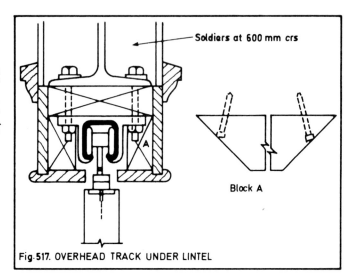

Fig.517. OVERHEAD TRACK UNDER LINTEL

Up and over doors

Up and over doors are very popular for garages. The weight of the doors, being compensated, makes them easy to operate and they slide up out of the way while taking up very little internal space. A number of different designs are made and patented.

Two types of up and over doors are shown here. In Fig. 518, the top of the door is carried back along tubular tracks on hinged brackets while a guide near the bottom on each side, attached to cables taken over pulleys to counterbalance weights, slides up vertical channels. A safety catch locks the door if the cable breaks.

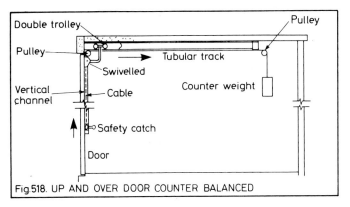

Fig.518. UP AND OVER DOOR COUNTER BALANCED

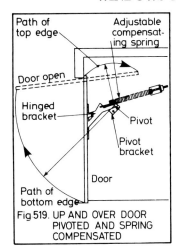

Fig.519. UP AND OVER DOOR PIVOTED AND SPRING COMPENSATED

In Fig. 519, the door is hinged on brackets fixed at each side and spring compensated. The top and bottom edges of the door follow the paths shown.

Generally speaking, all the mechanisms act separately either side of the door and are not inter-related. It is therefore necessary that they should be free-running and uniformly balanced.

Weather-proofing is provided in various ways, but commonly by an overlapping fillet nailed to each edge of the door on the inside at the top where it swings back and to the frame at the bottom on the inside where the door swings out. This is the same principle as the stops on pivot-hung sashes.

117

CHAPTER 10

Internal Fitments

The installation of built-in fitments to new buildings largely involves selecting suitable factory-made units, levelling them and setting them into place. The following chapter, however, deals with this subject more from the point of view of conversion and modernisation of older buildings, where the one-off required makes the built in situ item a viable proposition.

Built-in cupboards

Many old cottages ripe for modernisation, although solidly-built, seem to have been erected with the minimum use of the plumb-bob and spirit-level. They often have deep recesses in thick walls which can economically form the back and sides, of a wardrobe or cupboard. This leaves only a front gate frame and top, and shelves or drawers to be fitted with some adjustments to meet the unevenness of the main structure.

Figs. 520 to 526 give details of a wardrobe partly built-in and fitted between floor and ceiling. From the elevation

(Fig. 520), it will be seen that a panelled door is shown to a height of about 1.8 m, with a separate cupboard above for the storage of items not in general use.

The door to the cupboard would be fitted with a stay and catch which locks and releases on alternate pushups. The main cupboard space leaves sufficient room for a hat shelf at the top with coats and dresses below. These are supported by a stout rod or metal tube attached by brackets to the underside of the shelf.

The shelves and cupboard bottoms are carried on cleats fixed to the wall with nails into plugs, or by using a

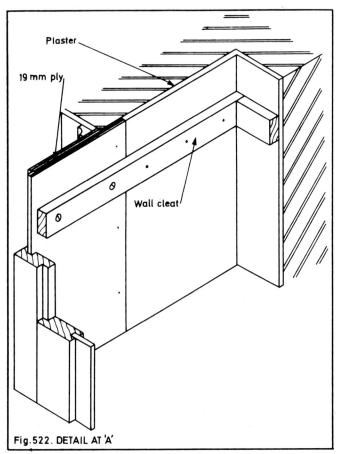

Fig.522. DETAIL AT 'A'

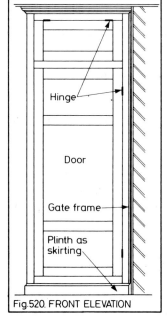

Fig.520. FRONT ELEVATION

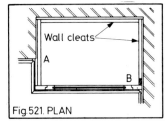

Fig.521. PLAN

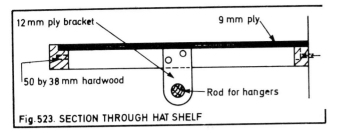

Fig.523. SECTION THROUGH HAT SHELF

12 mm ply bracket 9 mm ply,
50 by 38 mm hardwood Rod for hangers

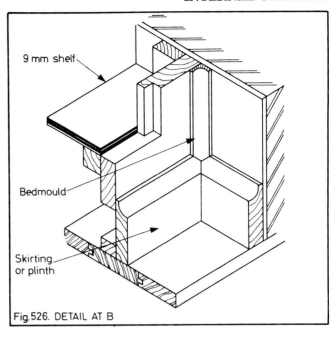

9 mm shelf

Bedmould

Skirting or plinth

Fig.526. DETAIL AT B

cartridge gun. Assuming a certain amount of inaccuracy, the front gate frame should be positioned first and the other items set to it.

To enable clothes on hangers to be placed transversely in the wardrobe, the front gate frame should be at least 560 mm deep (from front to back) which means the wardrobe may have to extend beyond the recess. Fig. 521 shows how this is accomplished; Fig. 522 giving a pictorial view.

The wardrobe is extended the required amount from the wall by a 19 mm plywood panel tongued to the front gate frame and rebated to width and depth as may be necessary to bring it flush with the plasterwork. A bed mould covers the internal angle joint.

The doors are rebated to the gate frame at the tops and sides as a precaution against dust and the entry of moths. The hat shelf (Fig. 523) may be framed up with 9 mm plywood supported by longitudinal and cross members. The front rail should be deep enough to carry its share of the not inconsiderable weight of clothing from the loaded rod or tube fixed to the shelf by brackets.

In Fig. 524, which is a part vertical section, the top is shown finished with a cornice. The lower mould of the cornice continues around the end and down the vertical internal angles to form a cover mould against the plaster. It mitres with the skirting of which the top member also matches. This is further illustrated in Figs. 525 and 526. Cupboard bottoms are supported at the front by cleats nailed to the insides of the relevant rails.

Drawers

Built-in units containing drawers can be more of a problem as, irrespective of any irregularity of the construction, both the drawers and the framing to receive them must be accurately made square, level and parallel in widths in all directions.

The components necessary for the fitting and smooth running of drawers are:

1. The rectangular framed opening to receive the drawer front.
2. The runners supporting the drawers. These must be dead level and flush with the framing.
3. The kickers over the tops of the drawers which prevent them from tilting when fully open. Where there is more than one drawer in depth, the runner of one becomes the kicker of the one below.
4. The guides which are square to the front rail keep the drawer straight and prevent it from jamming in the opening.

The drawer will probably be prepared in the shop and it will not be discussed here.

In order to ensure that these members are accurately fitted, the front gate frame and return end taken back to the recess are levelled, plumbed and fixed first and then a similar frame, with rails the runner thickness lower, is levelled and squared back from it. The runners are tongued to the front rails.

Kitchen dressers

Figs. 527 to 533 are drawings related to a kitchen dresser incorporating these details. Fig. 527, which is the front elevation, shows the bottom unit which has a worktop with drawers and cupboards under and an upper unit separately fitted between walls and carried on cleats. Fig. 528 shows the drawer in position on the left; and runners and guides to receive it on the right. Fig. 529 illustrates these details further in section.

Fig. 530 is a pictorial sketch giving details of the return end with the runner tongued into the front rail. Fig. 531

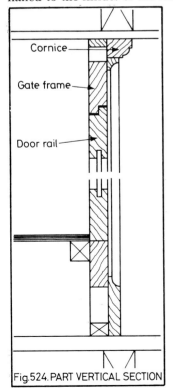

Cornice

Gate frame

Door rail

Fig.524.PART VERTICAL SECTION

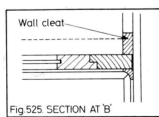

Wall cleat

Fig.525. SECTION AT 'B'

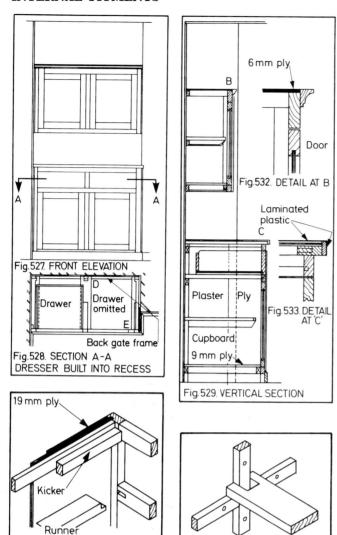

Fig.527. FRONT ELEVATION

Fig.528. SECTION A-A
DRESSER BUILT INTO RECESS

Fig.529. VERTICAL SECTION

6 mm ply

Fig.532. DETAIL AT B

Laminated plastic

Fig.533. DETAIL AT 'C'

Plaster Ply

Cupboard
9 mm ply

19 mm ply

Kicker

Runner

Fig.530. DETAIL AT 'E'

Fig.531. DETAIL AT 'D'

shows the runner supported by the rail of the gate nailed to the back of the recess. A clearance in the depth of the notch allows for unevenness in the depth of the recess.

The support for the bottom and shelf is obtained from the front and back gate and from cleats nailed to the wall between them. Enlarged details in Figs. 532 and 533 illustrate the finish to the dresser top and the build-up of the working top with 9 mm ply glued to suitable framing and faced and edged with laminated plastic.

Collapsible working surfaces

Where the working space in a kitchen is restricted, it may be convenient to provide temporary tables or tops which can be moved out of the way when not needed. In Figs. 534 and 535, sectional details are shown of a flap which folds up to enclose the free space above a dresser working top and opens down to give a wide, flat surface suitable for the preparation of foods, etc.

A wide rebate on the bottom rail of the framed flap fits a matching rebate on the inner flap resulting in a through level surface (Fig. 535). When the flap is closed, the projection of the rebated part below the hinge closes the gap between the inner and outer flaps to give a neat and dust-proof finish (Fig. 534). Note the bearers necessary to support the free edges.

The use of a continuous piano-hinge prevents the formation of a dust and dirt collecting groove at the joint. The hinge may be kept flush if the laminated plastic facing is bevelled at 45 deg. on the meeting edges.

An alternative means of forming an extension is from a flap which slides out from under the outer hinged member of the dresser top, as shown in Figs. 536 and 537. The outer flap, being hinged, lifts up and allows the extension to be pulled forward.

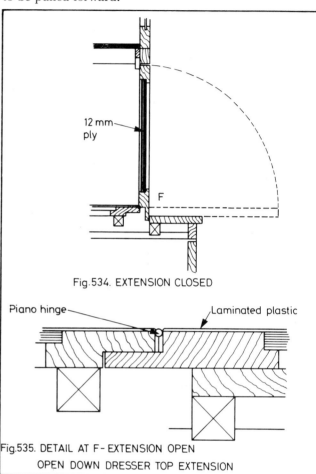

12 mm ply

Fig.534. EXTENSION CLOSED

Piano hinge

Laminated plastic

Fig.535. DETAIL AT F - EXTENSION OPEN
OPEN DOWN DRESSER TOP EXTENSION

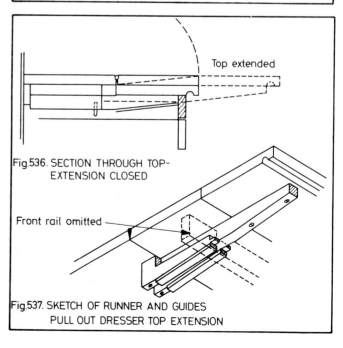

Top extended

Fig.536. SECTION THROUGH TOP-
EXTENSION CLOSED

Front rail omitted

Fig.537. SKETCH OF RUNNER AND GUIDES
PULL OUT DRESSER TOP EXTENSION

The extension is fixed to two slides or runners, the bottom edges of which are tapered to a pitch of the flap width by its thickness. The necessary parallel movement forward of the flap is controlled by suitable tapered guides, two to each runner, which are screwed to the inner flap and to the front rail of the dresser. Stops are formed by dowels fitted into the runners allowing about 25 mm excess movement so that the edge of the hinged flap may be reached for lifting to close.

Another method of forming a demountable working top or breakfast bar is by means of a folding wall table. The top hinged to a wall frame is supported by brackets which may be folded in to allow the top to be lowered when not needed.

The common type of bracket with the cantilever end supported by a brace is, in my opinion, ugly. It also has many crevasses to collect dust and, in the detail given, has been replaced by shaped 9 mm plywood.

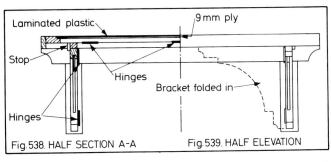

Fig.538. HALF SECTION A-A   Fig.539. HALF ELEVATION

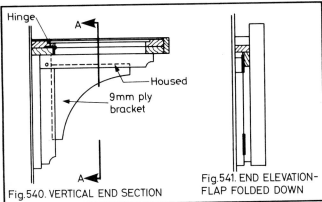

Fig.540. VERTICAL END SECTION   Fig.541. END ELEVATION-FLAP FOLDED DOWN

The flap itself is framed with solid rails and 9 mm ply faced with laminated plastic. The front edge needs to be stiffened sufficiently to avoid sagging between supports in use. The top is hinged to a rail placed underneath it so that, when in the horizontal position, no joint is seen on the surface. Figs. 538 to 541 illustrate the construction and are self-explanatory.

Serving hatches

Serving hatches between dining-room and kitchen are common to most domestic buildings and are most simply formed with single or double folding doors. These can sometimes be a nuisance and the best type are undoubtedly those which slide vertically to open. An example is shown in Figs. 542 to 545.

Some consideration has been given to finish and design and general appearance. The opening is fitted with a lining, as for an internal door, with an architrave mould covering the plaster joint on the inner face. The **door** slides

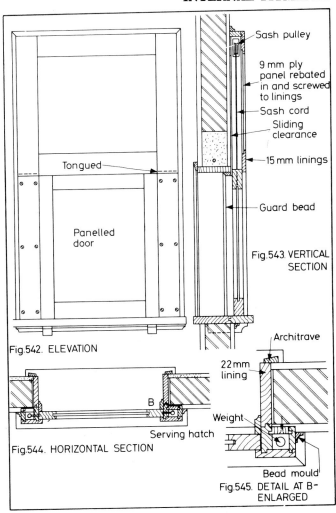

Fig.542. ELEVATION

Fig.543. VERTICAL SECTION

Fig.544. HORIZONTAL SECTION

Fig.545. DETAIL AT B- ENLARGED

between pulley stiles which must be twice its height. It must be counterbalanced, either by means of spiral balances, or by weights and pulleys. The latter, although requiring more complicated construction, appear generally to work more smoothly.

The base-board, or sill, has to project beyond the opening on both sides of the wall. For this reason, it is shown jointed in two parts, fitted from either wall face, and glued and jointed into position. It must project beyond the casings on the working side and is shown supported by brackets. The outer casings are tongued together, the panel being fitted into rebates. The lower parts of the outer casings are removable for renewal of cords.

The pulley stiles, which are a continuation of the vertical inner linings to the opening, are boxed together with the casings and finish against a top member. These details will be made evident from a general study of Figs. 542 to 545.

Pipe cases

The rest of the drawings illustrate minor fitments common to domestic buildings. Fig. 546 shows casings to conceal vertical service pipes. They may be framed and glued together, or dry rebated to vertical wall pieces with round head screw fixings for easy removal if needed. Where a stop-cock is fitted this, if concealed, should be made accessible through a hinged door (Fig. 547).

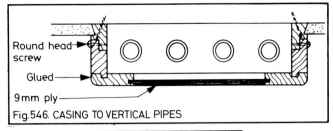

Fig.546. CASING TO VERTICAL PIPES

Round head screw

Glued

9mm ply

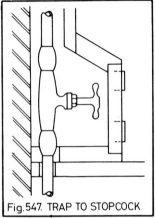

Fig.547. TRAP TO STOPCOCK

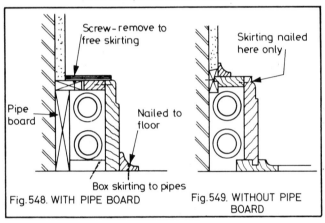

Screw – remove to free skirting

Skirting nailed here only

Pipe board

Nailed to floor

Box skirting to pipes

Fig.548. WITH PIPE BOARD

Fig.549. WITHOUT PIPE BOARD

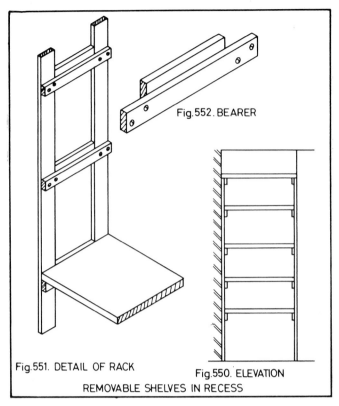

Fig.552. BEARER

Fig.551. DETAIL OF RACK

Fig.550. ELEVATION

REMOVABLE SHELVES IN RECESS

Pipes in bathrooms and kitchens placed at floor level may be boxed behind the skirting. They must be reasonably accessible in the event of a leakage or pipe burst.

In Fig. 548, it is assumed the pipes are fixed to a board which may then be used to support the boxing, the skirting only being held by a few screws through the plywood capping. In Fig. 549 it is presumed the carpenter is given the problem of casing in pipes already fixed close to the wall. Here grooved battens nailed to the wall and floor carry the capping and skirting, the skirting being fixed only by a few nails through the top edge.

Shelving

Shelves fitted in recesses can, of course, be carried by cleats nailed direct to the plugged wall; but by use of simple ladder racks, they may be fitted in with little or no fixing as shown in Figs. 550 to 552. Fig. 550 shows the shelves in position carried by the ladders. Fig. 551 is a sketch of part of one of the latter with the position of a shelf. Fig. 552 shows a sketch of one of the bearers.

By rebating each bearer as shown in Fig. 552, the shelves do not have to be notched and may be slid straight into position. This also avoids the formation of little pockets for dust and insects behind the bearer.

Where shelves up to about 300 mm wide have to be fitted against a wall, they can be most easily supported by steel brackets as in Fig. 553. They are best fixed to continuous vertical battens which will bridge local variations in the plaster and keep the brackets plumb and square. If the battens are housed out for each shelf as shown, this will conceal a bad fit in the shelf notch due to any irregularity in the wall surface and will hold the shelf against the screws driven through the brackets.

Where shelves have to be returned at right-angles, the joint is best formed with a flat splay as in Fig. 554. It makes a better job if this is glued with one of the modified pva glues and held to set with a G-cramp.

Where wide shelves are needed, it is better to make up timber brackets to support them; but if more than one shelf has to be supported, all the brackets in one vertical row should be formed as part of one continuous wall unit as in Fig. 555. This shows one shelf support in position and provision for the next one below. Joint details are also shown.

Traditionally, the brace to each bearer is made flat to coincide in width with the other members. It may be a

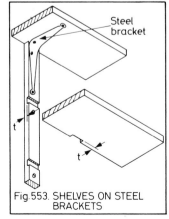

Steel bracket

t

t

Fig.553. SHELVES ON STEEL BRACKETS

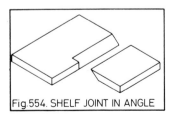

Fig.554. SHELF JOINT IN ANGLE

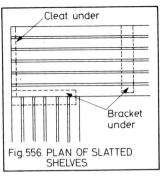

Fig. 556. PLAN OF SLATTED SHELVES

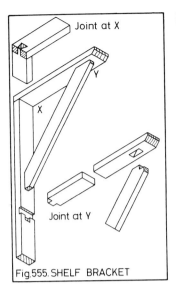

Joint at X

Joint at Y

Fig. 555. SHELF BRACKET

little stronger than that sketched; but, in my opinion, the bracket shown is more positive in construction, much neater to look at and takes up less space in the length of the shelf. If, for any reason, the shelf is not level e.g. wall out of plumb; this may be corrected.

Wide shelves are more cheaply formed with open battens, as shown in Fig. 556, then with solid timber. However they have the disadvantage that they form many dust traps. A better method may be to use plywood or chipboard suitably supported by longitudinal members as in Fig. 523.

Proprietary panels

Ordinary chipboard may be used in the same way as plywood for panels, facings, shelves, etc., although edges should be protected. They may be lipped with narrow strips of wood, preferably tongued into grooves in the chipboard edges.

Manufacturers have, however, produced a wide variety of sizes of component panels which are surfaced both on sides and edges with coloured melamine or wood veneer. These are supplied in a number of standard widths, in lengths of up to 2.4 m. They are primarily designed to be cut to length only, and then used for all parts of a fitment; i.e. doors, shelves, sides, tops, bottoms and drawer-fronts.

Complete sets of special tools and accessories are available for converting the panels into various types of furniture and fitments, all constructions being completely box-like in design. Special tools are:

wood boring bits with depth gauges for dowel holes;
centre-points for inserting in one lot of holes and marking the centres for matching holes in dowelled joints;

jigs for holding the panels at right-angles to each other and for positioning the bits for matching holes for dowels in both components to be jointed together at right-angles;
hard steel knives for scoring the melamine surface before cutting; and
edge trimmers for cutting the veneer or laminate edge strips.

Accessories are:

cupboard catches for screwing to door and carcase,
chipboard plugs for use with screws to give a specially firm anchorage for handles, etc.;
knock down corner joints, made in two parts and screwed separately to each jointed unit which may then be assembled with bolts through the fittings;
shelf supports, comprising sockets fitted into drilled holes to receive special studs to support the shelves;
rigid corner block screwed into each panel jointed at the angle with cover plates over the screws;
screw covers, the screw is also driven through a plastic washer on to which a cap is snapped to conceal the screw; and
special hinges screwed to the inside face near the edge of door and carcase.

Drawer material consists of long lengths of hollow rectangular units which incorporate a groove for a runner and another suitably placed to receive the drawer bottom.

The units are squared off to length and jointed at the angle with two-way brackets designed to drive tight into the hollow ends. The bottoms on melamine-surfaced hardboard, or other suitable thin material, are inserted at the same time.

The fronts may be applied afterwards or may form one side of the drawer using a special single bracket which is first screwed, to the drawer front. The drawer runs on wood or plastic runners accommodated in the groove in each boxed side. The panels will safely span about 600 mm for load-bearing shelves, bottoms, etc. Over this span, some extra timber bearers will be needed.

In applying the system to the maximum economic advantage, cupboards, wardrobes and etc., should be planned to accommodate the available component widths. Any difference, such as where the unit has to fit into a recess, should be made up in the member, such as a drawer carcase, where the components are used horizontally and may be cut to length. The system can, of course, be incorporated into normal joinery construction with greater flexibility.

CHAPTER 11

Timber Buildings

Small buildings constructed mainly in timber may be placed in three categories:

1. Temporary buildings: sheds, offices and stores on a building site are temporary in the sense that they will eventually be dismantled but they will be erected again elsewhere.
2. Permanent buildings, workshops and sheds for the storage of tools and small offices on private residences, farms, car parks, etc.
3. Comparatively long buildings with large doors used as garages or for storing farm equipment for which some extra strengthening is needed.

TEMPORARY BUILDINGS

Temporary buildings on site may vary greatly in size, layout and construction. On a large site, there may be a suite of offices for the administration staff, a canteen, and separate stores for fittings, cement, etc.

On a small contract, there may be just one building for offices and general use, plus another for cement storage; although where a clerk of works is in permanent residence, he must be provided with a separate office.

The accommodation needed for a clerk of works and general foreman are similar and comprise a long desk for making and layout of drawings, a plan chest, a level shelf or table where materials may be tested and weighed, and some shelves for storage of especially valuable equipment and fittings.

A wash basin will be needed and, in the winter, some safe form of heating. Windows should be sufficient in size and number to provide light for working and the buildings should be positioned so that they give a good view of the site. However sufficient wall space is also needed for charts commonly consulted. The door should be fitted with a letter plate, and a pay hatch may be needed in the foreman's office.

Figs. 557 to 560 are the elevation, horizontal section, vertical section, and roof plan of a typical clerk of works office. From Fig. 558, it will be seen that the walls are in

three sections on the one side; and two on the other with windows. The end with the doorway is in one piece; although it could have been in two with a joint midway over the door head which may be removable.

The outsides of the walls are shown covered with vertical rebated boarding. Matchboarding is commonly used, but the tongue is fragile. Under exposed conditions, it is likely to break; and with water trapped in the groove is

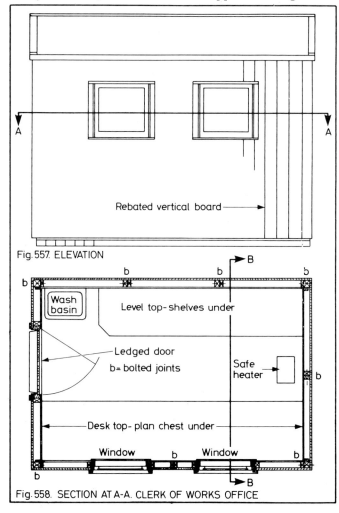

Fig. 557. ELEVATION

Fig. 558. SECTION AT A-A. CLERK OF WORKS OFFICE

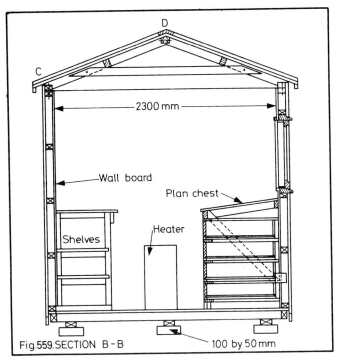

Fig.559. SECTION B-B

2300 mm

Wall board

Plan chest

Shelves

Heater

100 by 50 mm

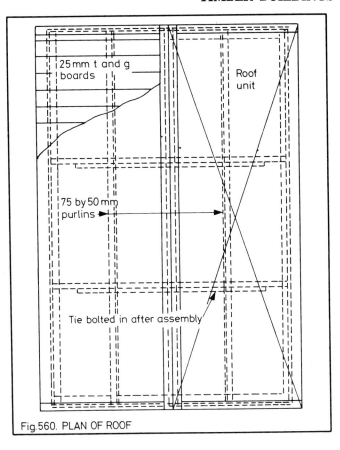

25 mm t and g boards

Roof unit

75 by 50 mm purlins

Tie bolted in after assembly

Fig.560. PLAN OF ROOF

more likely to decay. Alternatively, the walls could be faced with 6 mm plywood, external grade; but the edges should be protected with a cover fillet and preferably treated with preservative.

The jointing of the section is shown in Fig. 561. At the corners, the sides are bolted to the ends which have 75 by 75 mm corner posts. If the inside of the walls are lined with wallboard or hardboard, this will give a greater comfort to personnel and save on heating, as well as creating a better impression on visitors.

It is necessary to nail a 50 by 25 mm batten to the wall post to receive the end of the wall boarding. The joint is weather protected by taking the side boarding about 30 mm over the post. Pockets are necessary in the wall linings to give access to each bolt for assembly and dismantling. Cleats are screwed through the wall boarding to take the pocket pieces as shown. Intermediate joints between the wall sections are also shown, the inner and outer linings of the one section lapping the other. The overlapped stud may be reduced in size.

Doors and windows

In order to reduce costs on the job, it is common practice to allow the carcase framing itself to form the door and window frames; but in my opinion, it is better to form these as separate units, even if they are only 25 to 38 mm thick. The timber in these buildings, which are always neglected, will be continually subjected to the weather and the use of separate frames enables the necessary joints to doors and windows to be better waterproofed; thus adding many years to the life of the building.

Fig. 562 shows a horizontal section through a window jamb, taken to the face of the wall with a cover mould over the joint. A planted stop is used to form the rebate. Fig. 564 shows a section through the window head, which

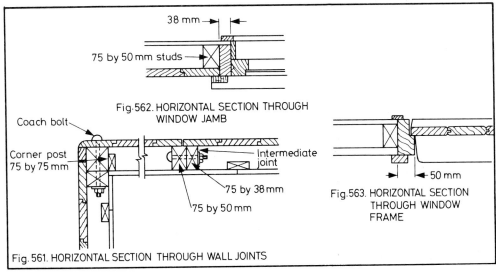

38 mm

75 by 50 mm studs

Fig.562. HORIZONTAL SECTION THROUGH WINDOW JAMB

Coach bolt

Corner post 75 by 75 mm

Intermediate joint

75 by 38 mm

75 by 50 mm

50 mm

Fig.563. HORIZONTAL SECTION THROUGH WINDOW FRAME

Fig.561. HORIZONTAL SECTION THROUGH WALL JOINTS

125

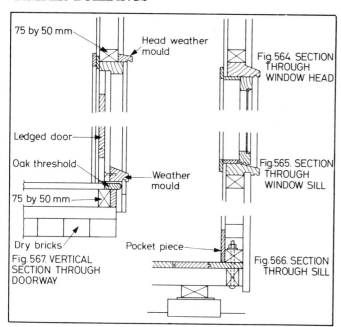

Fig. 567. VERTICAL SECTION THROUGH DOORWAY

Fig. 564. SECTION THROUGH WINDOW HEAD

Fig. 565. SECTION THROUGH WINDOW SILL

Fig. 566. SECTION THROUGH SILL

has a weather mould with a drip groove throwing the water clear of the window rebate.

The sill (Fig. 565) is weathered in the normal way, and the external lining is taken into a groove in the sill. The inside is finished with a planted stop and band mould.

Floors

The building has a wooden floor which could be in 1.2 m sections with 75 by 50 mm joists at 400 mm centres. A 75 by 38 mm header is nailed across the ends of the joists. The flooring is 22 mm tongued and grooved boarding.

It must be appreciated that the floors of site buildings may have to carry considerable weight, particularly storage sheds and cement huts. Therefore some care should be taken to provide suitable foundations. Fig. 559 shows header bricks resting on the compacted soil and carrying 100 by 50 mm plates. Alternatively, old sleepers could be used direct. If the building has to stay a long time, it is just as well to interpose some form of damp-course between the floor and the foundation.

It may be more convenient to bed the bricks or sleeper on a level layer of weak concrete. Any settlement in the foundations may cause the door to jamb.

Fig. 566 shows the wall section bolted to the floor either through the joist or side header. Pockets will be needed in the lining to give access to the bolts. Note that the wall boarding extends below the floor to form a drip.

Fig. 563 shows a horizontal section through the door frame and edge of the door which is commonly ledged, or ledged and braced. A weather mould is fixed to the bottom of the door. This may be considered an unnecessary refinement, but it will keep the threshold dry and reduce the risk of decay. Other details are shown in the vertical section in Fig. 567.

A head weather mould keeps the top edge of the door dry while it can also be seen how the weather mould to the bottom of the door throws the water clear of the step.

The roof

The roof (Figs. 559 and 560) is made in two parts coming together at the ridge where the joint is covered by a ridge board unit. Each part of the roof is framed up with rafters, plate and ridge and purlins cut midway between the ridge and plate. Fig. 568 shows further details of construction.

Purlins are stub tenoned into the rafters. The roof may be boarded with plywood or chipboard. The weather-proofing is a double layer of ruberoid, preferably put on with mastic as this is less likely to be torn. The bottom edge is formed with a drip secured to a fillet screwed to the edge of the sheeting. A 300 mm strip is nailed to the fillet and folded back over the nails to line up with the underfelt (Fig. 569).

Figs. 570 and 571 show the arrangement of studding to the walls with alternative methods of strutting. The struts should go from corner to corner of the main rectangles and intermediate members cut against them.

Framing

In arranging the framing, due consideration should be given to providing supports to the joints with intermediates at suitable spacing for both inside and outside linings. Thus, in Fig. 570 horizontals must be provided at suitable intervals to take the vertical external boards, while inside, provision must be made for the edges of 1.2 m wide wall boarding.

The actual jointing of members may be by stub tenon, halving or notching; but generally, nailed butt joints are considered sufficient. Nails driven through members near the ends should have the holes predrilled to 4/5ths of the nail shank diameter.

In constructing portable buildings, it must be remembered that they are likely to come under the greatest strain when mishandled under transport. To withstand this, bracing becomes very necessary. Fig. 571 shows the end framing with the door opening. A weak spot occurs at

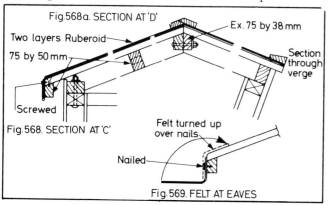

Fig. 568a. SECTION AT 'D'

Fig. 568. SECTION AT 'C'

Fig. 569. FELT AT EAVES

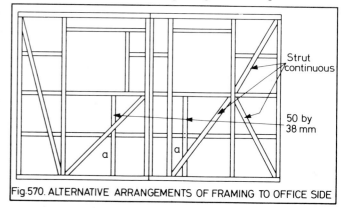

Fig. 570. ALTERNATIVE ARRANGEMENTS OF FRAMING TO OFFICE SIDE

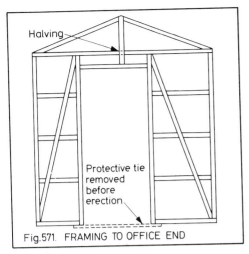

Fig.571. FRAMING TO OFFICE END

the top. This weakness is reduced by using a halving joint as shown. A cleat across the door opening at the bottom will also give temporary support during transport.

Internal fittings

The shelf units (Fig. 559) are carried on ladder frames screwed to the wall through the back stiles. The edge of the top is supported by a front rail. The desk top is framed up of plywood or chipboard with front and back rails and cross bearers at about 600 mm centres. It is screwed to the wall through the back rail and supported by struts carried by brackets at the bottom of the wall section. The plan chest slides under the desk top.

Special consideration should be given to the safety of the heater. It should, preferably, be fixed by a bracket to, but away from, the wall which should be shielded with an asbestos or metal plate.

Assembly

Figs. 572 to 574 illustrate a method of assembling timber buildings of any convenient length using narrow standard units. The units are all interchangeable and can be glazed or otherwise to give the amount of light required. It is important that they should all be built in a jig with bolt holes set exactly to a templet.

Finishings can be the same as in the previous example. The roof units are formed with ex-225 by 38 mm joists tapered both ways to fall and the same width as the sections.

A fascia is nailed to the joist ends, extending below the bottom ends of the joists to form a drip. To straighten the top edge formed by the wall units and eliminate any unwanted lateral flexibility, a 225 by 25 mm board is screwed to the heads as seen in Fig. 574 and is held up against the joists by a cleat nailed inside the fascia.

The joists are lined with wall board to form a ceiling, a gap being left to enable the wide board to fit up and slide behind the fascia cleat. Any joint in the soffit board should be midway on a section.

Garages

It is common practice to build garages of noncombustible material because of the obvious fire risk. The Building Regulations (Part E18) only lay down the proviso, which in non-legal phraseology means, that if a garage is less than 2 m from a building or a boundary, that part which is within that distance from the building or boundary must be lined with Class O surface material on the inside and a non-combustible material on the outside. There is nothing against the building being framed in timber. It would be wise, however, always to use incombustible materials on inside and outside surfaces.

The structural problems in building a timber-framed garage are its lateral weakness at the top, especially when

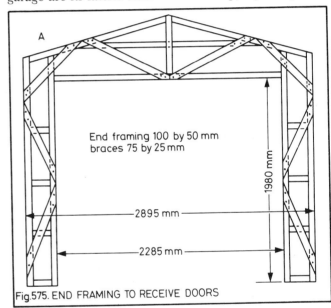

Fig.575. END FRAMING TO RECEIVE DOORS

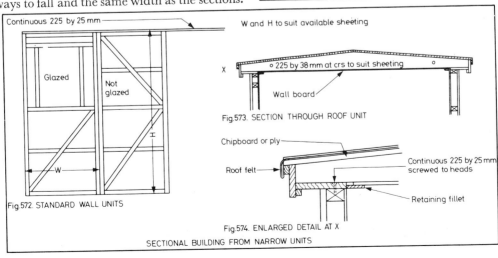

Fig.572. STANDARD WALL UNITS

Fig.573. SECTION THROUGH ROOF UNIT

Fig.574. ENLARGED DETAIL AT X

SECTIONAL BUILDING FROM NARROW UNITS

designed to take two cars in depth, and the weakness at the open end due to the necessary full width open doorway.

Fig. 575 shows the framing to a garage with a doorway which leaves about 300 mm width of building on either side. The timber necessary to provide nailing for the outer sheeting is assembled and then diagonal bracing is introduced in such a way as to give connected triangulation to both head and jamb framing.

In order to give level nailing surfaces, without undue weakness, 100 by 50 mm studding is used and 75 by 25 mm bracing is housed flush into it. The construction is illustrated pictorially in Fig. 576. The side walls are assembled in units about 1.2 m wide and suitably braced to give rigidity. Bear in mind that the noncombustible sheeting will do little to add strength to them (Figs. 577 and 578).

It is assumed that the roof will be sheeted with corrugated asbestos cement for which purlins are required at the correct spacing for fixing and support. They are carried by rafters which, in turn, are attached to vertical posts between consecutive wall sections (Figs. 578 and 579).

The wall sections are 75 mm deep. The posts being 125 by 38 mm stand inside the walls by 50 mm. They also stand above the wall sections and the rafters are nailed to them. Ceiling ties are nailed to the rafters in line with the posts. To give rigidity and lateral stiffness 75 by 25 mm knee braces are nailed to the tie and the exposed 50 mm

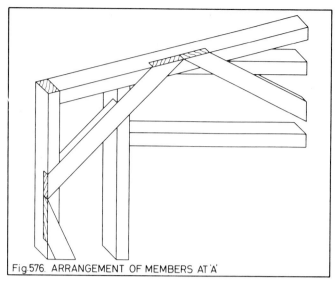

Fig.576. ARRANGEMENT OF MEMBERS AT 'A'

of the posts. These details are shown in the end section in Fig. 580 and in the pictorial sketch in Fig. 581.

The garage would have a concrete floor which could have with advantage a 100 mm riser to which the sills of the garage walls could be bolted with damp-proofing in between; thus reducing the risk of decay.

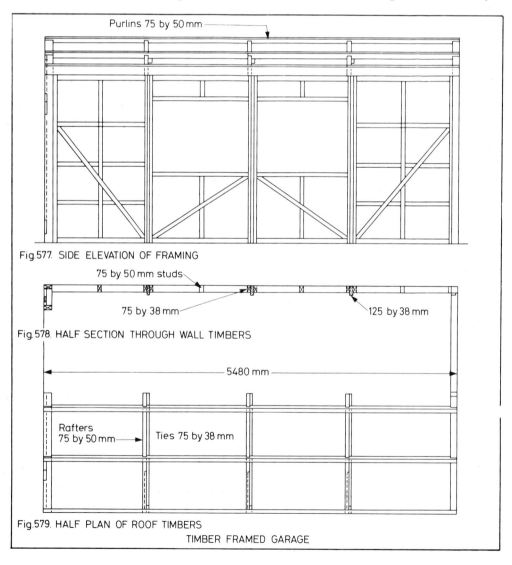

Purlins 75 by 50 mm

Fig.577. SIDE ELEVATION OF FRAMING

75 by 50 mm studs

75 by 38 mm

125 by 38 mm

Fig.578. HALF SECTION THROUGH WALL TIMBERS

5480 mm

Rafters 75 by 50 mm

Ties 75 by 38 mm

Fig.579. HALF PLAN OF ROOF TIMBERS

TIMBER FRAMED GARAGE

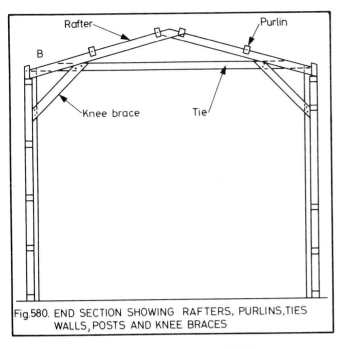

Fig.580. END SECTION SHOWING RAFTERS, PURLINS, TIES WALLS, POSTS AND KNEE BRACES

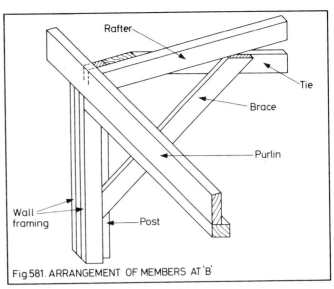

Fig.581. ARRANGEMENT OF MEMBERS AT 'B'

DOMESTIC BUILDINGS

Timber-frame domestic buildings vary from those of traditional brick construction mainly in the walls and partitions. Floors and roofs are normally of timber in either case. It is therefore only with regard to walls that the Building Regulations have to be given special consideration. This will have to be done under four headings:

1. thermal insulation;
2. structural fire precautions;
3. structural stability; and
4. sound insulation, in the case of separating walls.

THERMAL INSULATION: presents no problems. There are a wide variety of possibilities available, from the use of finishings in the form of external claddings and internal linings, with the choices of loose fill; flexible insulation, such as quilting or blanket; rigid insulation of cork, or fibre, or plastic; and reflective metals within the cavities formed by the framing.

STRUCTURAL FIRE PRECAUTIONS: are prohibitive to timber construction above certain sizes. They also become more stringent for walls erected close to boundaries or existing buildings.

STRUCTURAL STABILITY: of timber walls is increased above the simple load-bearing strength of vertical studs by the stiffness given by diagonal struts or braces, when present, and internal and external linings; as well as additional support from return walls and intermediate partitions.

The Building Regulations 1985 give little information on this but state a general rule that timberwork shall comply with BS 5268 parts 2 and 3. However the standard methods of construction are likely to bring the work well within these requirements.

SOUND INSULATION: The double requirements of sound insulation and fire protection in separating walls are met by using hollow timber construction with twin isolated timber frames, faced with a 30 mm layer of plasterboard so that there is an internal gap of at least 200 mm in which is suspended a 25 mm layer of unfaced mineral fibre quilt with a density of 12 kg/m^2. The Building Regulations for chimneys, flues, hearths and fireplace recesses (Approved Document J) can be applied directly to timber studding. Where fixings to the brickwork are needed, this may have to be extended to give a safe cover.

Foundations

In all timber constructions, protection against damp either from wind and rain, capillary action or condensation is a necessity; great care must be taken in detailing and following up to ensure that this takes place. It goes without saying that all foundation work below the damp course must be in brick or concrete.

The set out of the foundation work for timber frame construction is less complex than for traditional brick buildings. It is confined to perimeter walls, foundations for fire-places and intermediate sleeper walls. The latter are to reduce the span of floor joists and permit a subsequent reduction in their cross-sectional area. Timber plates are laid on a damp-proof course on the walls, being bolted around the perimeter at intervals of about 1.2 m. Joists are laid and nailed to these at suitable spacings over the whole of the floor area and then completely boarded over (clear of any fireplaces) with 12.7 mm plywood or 18 mm diagonal boarded sub-floor.

On this surface, if the work is being done on the site, the walls and partitions are set out, assembled and ultimately erected. The space within the floor joists is totally enclosed by heading joists laid continuously on the side plates and nailed to the ends of the floor joists. Fig. 582 shows part of a floor to a timber-framed building with foundation walls, sleeper walls, sill plates and joists with some 2.4 by 1.2 m sheets of 12.7 mm WBP (weather and boil proof) plywood sub-flooring in position. It should be laid with the external grain laid parallel to the span to give the greater degree of stiffness. Sleeper walls are shown spaced at 1.8 m centres. This floor will have to carry all the partition walls and any loads transmitted to them. It may be convenient to arrange the sleeper walls to come directly under any heavily loaded partition walls.

129

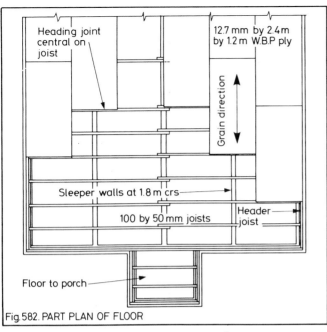

Fig. 582. PART PLAN OF FLOOR

(labels in figure: Heading joint central on joist; 12.7 mm by 2.4 m by 1.2 m W.B.P ply; Grain direction; Sleeper walls at 1.8 m crs; 100 by 50 mm joists; Header joist; Floor to porch)

Methods of timber-frame construction

Figs. 583 and 584 are pictoral sketches showing alternative methods of timber-frame construction. In Fig. 583, in what is known as balloon frame construction, the vertical studs rest directly on the wall plate or sill and are nailed to the sill and to the sides of the joists. The studs extend to the full height of the wall and the first floor joists are carried on 100 by 25 mm ribbons housed flush into the studs.

For structural fire precautions, and to meet the Building Regulation requirements, all cavities formed within walls, floors and roofs must be isolated by suitable 38 mm timber fire stops at their extremities or at certain specified distances.

To meet those requirements, the floor cavities must be sealed off at the level of the tops of the floor joists and at both edges of the first floor joists.

The studs must be spaced at centres which are a factor of the width of the linings, commonly 400 mm to 600 mm. If the studs are to be sheeted externally with plywood, this will provide the stiffness to resist distortion from lateral stresses, i.e. wind, etc.; when well nailed in position. Otherwise, if the external lining is of a softer material such as plasterboard or wallboard, some diagonal bracing housed in flush becomes necessary as shown in the sketch.

Fig. 584 illustrates timber framing in platform construction. By this method, each floor is complete in itself. The ground floor and foundations are laid and sheeted over right to the edges. The exterior walls and partitions are then either supplied made-up or assembled on the floor. The former are sheeted or braced to keep them square. They are then lifted into position, using temporary strutting as may be necessary, and nailed together.

The first floor joists are laid and sheeted all over, as before, and the whole procedure is repeated for walls and partitions. The building can then be roofed over, generally, using preassembled trussed rafter construction, so leaving the rest of the work to be done in the dry.

Protection against damp

Fig. 585 is a section through the perimeter wall of the foundations and illustrates the precautions necessary to protect the timber against damp. Building Regulation requirements as to footings and oversite concrete have to be met. The sill plate must be held down using suitable anchor bolts. If the wall is brickwork, the bolt is passed through a horizontal steel plate bedded in the joint.

The damp course must extend beyond the sill to protect

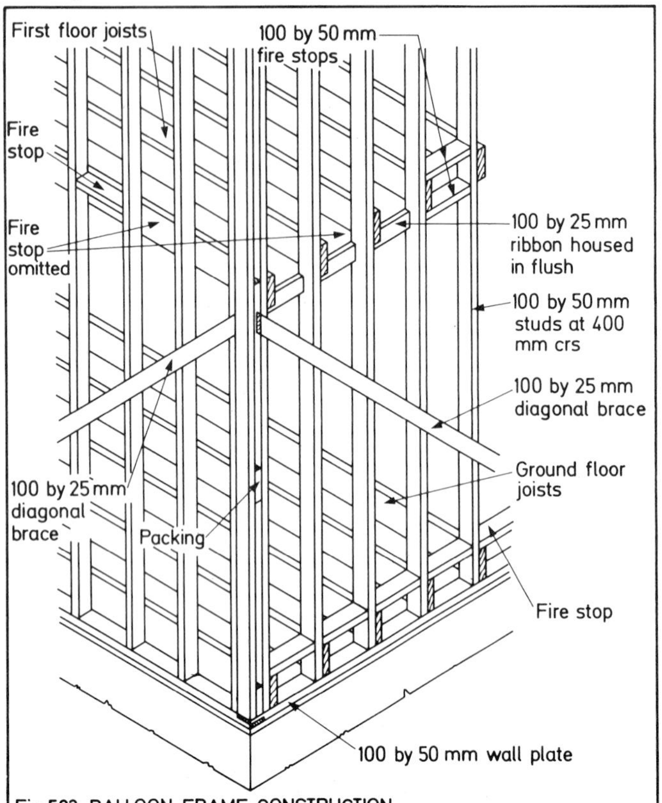

Fig. 583. BALLOON FRAME CONSTRUCTION

(labels in figure: First floor joists; 100 by 50 mm fire stops; Fire stop; Fire stop omitted; 100 by 25 mm ribbon housed in flush; 100 by 50 mm studs at 400 mm crs; 100 by 25 mm diagonal brace; Ground floor joists; 100 by 25 mm diagonal brace; Packing; Fire stop; 100 by 50 mm wall plate)

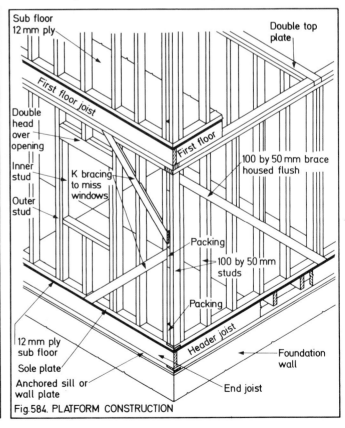

Fig. 584. PLATFORM CONSTRUCTION

(labels in figure: Sub floor 12 mm ply; Double top plate; Double head over opening; First floor joist; First floor; 100 by 50 mm brace housed flush; Inner stud; K bracing to miss windows; Outer stud; Packing; 100 by 50 mm studs; Packing; 12 mm ply sub floor; Sole plate; Anchored sill or wall plate; Header joist; Foundation wall; End joist)

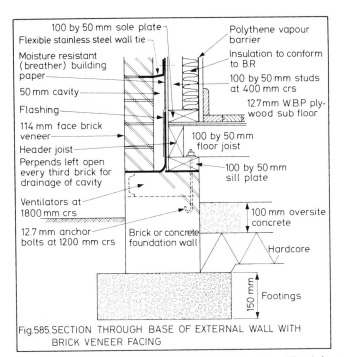

Fig.585.SECTION THROUGH BASE OF EXTERNAL WALL WITH BRICK VENEER FACING

Labels in figure:
100 by 50 mm sole plate
Flexible stainless steel wall tie
Moisture resistant (breather) building paper
50 mm cavity
Flashing
114 mm face brick veneer
Header joist
Perpends left open every third brick for drainage of cavity
Ventilators at 1800 mm crs
12.7 mm anchor bolts at 1200 mm crs
Polythene vapour barrier
Insulation to conform to B.R
100 by 50 mm studs at 400 mm crs
12.7 mm W.B.P plywood sub floor
100 by 50 mm floor joist
100 by 50 mm sill plate
Brick or concrete foundation wall
100 mm oversite concrete
Hardcore
150 mm Footings

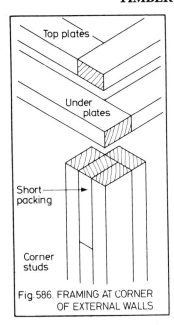

Fig.586. FRAMING AT CORNER OF EXTERNAL WALLS

Labels: Top plates, Under plates, Short packing, Corner studs

the edge of the plywood (usually 8 mm thick). The joists and sub-floor are laid next, and the wall frames erected. All the timber work must then be protected using metal or other suitable flashings extending above the floor and turned out to the edge of the foundation wall.

The timber-framed wall must also be made safe against the penetration of rainwater, assisted by wind, through the outer sheathing; and also against water vapour (always present in occupied buildings) passing through the inner linings and condensing within the colder cavities.

The inside of the walls are lined with a vapour barrier of polyethylene sheeting which is completely impervious. It is anticipated that, through practical imperfections, some vapour may still pass into the hollow framing; so the outside is lined with a moisture barrier or 'breather paper'. This keeps water out but allows vapour to pass through. Thus it allows any vapour within the cavities to escape when external conditions permit. These items are shown in the section. The moisture barrier will overlap the flashing at the base.

Strengthening the walls

There are various methods of finishing the outside of the walls, some of which will be shown later. In Fig. 585, the outside of the wall is faced with a 114 mm walls of brick-work with a 25 to 50 mm cavity, thus providing a non-combustible exterior and the appearance of a traditional brick house. The wall is anchored to the framing by metal flexible ties screwed through the plywood to the studs at intervals of about 900 mm horizontally and 450 mm vertically. The wall will not carry any weight and can actually be built after the timber structure is completed.

Some additional thermal insulation will be needed within the hollow construction as shown in the section. This can be one of a number of different types of materials to give the required 'U-value.

The use of elaborate joints in assembling the framing is not necessary. Well nailed butt joints, say two 100 mm wire nails into the end grain of each stud, are considered satisfactory. The strength will be provided by the external and

internal linings. Wherever internal angles are formed in the walls, provision has to be made for a secure nailing for the linings. This is achieved by introducing an additional stud.

In Fig. 586, the stud is blocked off the external corner to give a nailing in the return inner angle. In Fig. 587, at a tee-junction, two studs are spaced 50 mm apart and the stud of the return framing comes centrally on to them. In Fig. 587, the sheeting should first be nailed to the narrow edge.

To help give continuity along the length of the wall and provide extra strength to carry a loaded joist bearing on a mid-spacing, the heads of the framing are usually double

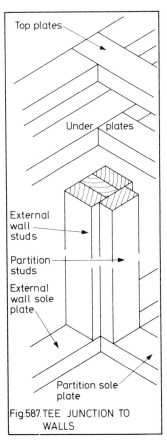

Fig.587. TEE JUNCTION TO WALLS

Labels: Top plates, Under plates, External wall studs, Partition studs, External wall sole plate, Partition sole plate

thicknesses of 50 by 100 mm timber. The lower member should be jointed over the top of a stud and the top ones, lapping at least 1.2 m should also be jointed directly over a support. Any angles should also be formed with lapped joints as shown.

The roof

Gable ends should be built up into separate triangular frames as in Fig. 588 and erected as part of the roofing operation. Overhanging verges will need some support; this is most easily obtained from cantilever bearers tied back to the first trussed rafter taken two rafter spaces in. Intermediate support to roofing should be given by short members cut in between the bearers. Additional short members may then be nailed to the end rafter and gable head to carry the soffit to the verge. These details are shown in Fig. 589.

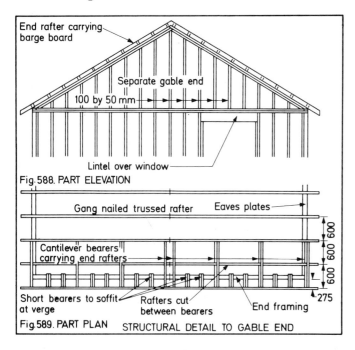

Fig. 588. PART ELEVATION

Fig. 589. PART PLAN   STRUCTURAL DETAIL TO GABLE END

The roof itself may be of standard construction but some extra lateral stiffness may be needed other than that provided by the roofing battens. This can be obtained by diagonal braces of, say, 100 by 15 mm nailed up under the rafters.

The finish at the eaves is straightforward, fixings being easily obtained. Fig. 590 shows a section at the eaves where the wall is faced with timber siding while Fig. 591 shows a finish against an external brick veneer. In America and Canada, where weather conditions are more severe, roofs are fully ventilated at the eaves and ridge; and the ceilings are heavily insulated so that the temperature of the roof space is approximate to that of the open air. This is to prevent the penetration of water, which could occur if melting snow at the upper part of the roof was allowed to form pools retained by frozen snow at gutter level.

On gable ends the finish at the verge is usually formed with a barge-board. The edges of the tiles or shingles are either notched over it or protected by a capping nailed to the top edge as in Fig. 592. Fig. 593 shows the lower part of the barge-board extended to cover the end of the fascia and soffit at the eaves.

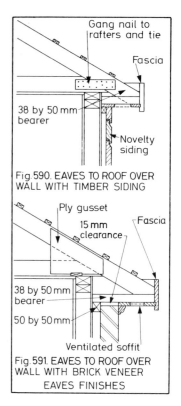

Fig. 590. EAVES TO ROOF OVER WALL WITH TIMBER SIDING

Fig. 591. EAVES TO ROOF OVER WALL WITH BRICK VENEER EAVES FINISHES

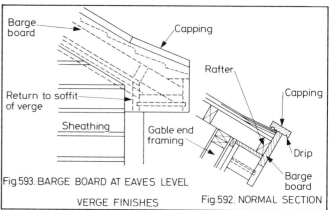

Fig. 593. BARGE BOARD AT EAVES LEVEL   VERGE FINISHES

Fig. 592. NORMAL SECTION

Finishing exterior walls

The design of the external facings or sheathing to the timber-framed wall requires careful consideration, if timber is used for this the cladding is likely to be continuously wet in winter weather and it is essential that the damp should not be allowed to penetrate to the plywood sheathing. This is best achieved by forming a cavity between the two, 25 mm thick vertical and horizontal battens nailed through the plywood sheathing at 400 to 600 mm centres carry horizontal and vertical timber siding. The battens are faced with dpc to prevent the moisture from crossing the cavity. The battens and the sidings if of softwood should be pressure impregnated with preservative. This construction is shown in further diagrams. Complete and tight fitting tongues and grooves should generally be avoided as, although they may appear to be almost watertight, they invite capillary action and in the winter may never be dry. Added to this, they tend to be fragile and the grooved edge may ultimately shrink and take the split-off tongue with it.

The bottom edge of the sheathing is vulnerable as it is the last area of timber to dry after a storm. It should

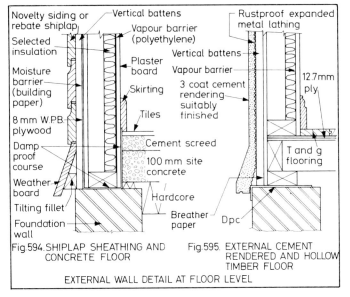

Fig.594. SHIPLAP SHEATHING AND CONCRETE FLOOR

Fig.595. EXTERNAL CEMENT RENDERED AND HOLLOW TIMBER FLOOR

EXTERNAL WALL DETAIL AT FLOOR LEVEL

Fig. 596 gives an alternative of horizontal feather edged boarding. As this is nailed on both edges, it should be in narrow widths to reduce individual shrinkages and to avoid splitting. It should be dry when fixed. One effective method of sheathing is by the use of double rows of Western red cedar shingles (Fig. 597). Duplication in the row with staggered joints ensures that all vertical joints are covered and there only needs to be a single vertical lap between rows. The inner shingles may be of an inferior quality as they are not seen. The outer members in each row are set 12 mm low giving a free ventilated edge to drip clear.

Fig. 598 gives two methods of vertical boarding. In the first, known as 'extended shiplap', each board is nailed only on its rebated edge, thus allowing the other edge full moisture movement. In the board and batten method, the boards are square edged and laid with open joints between which the nails holding the battens are driven. The boards themselves are held by one vertical row of nails through the middle, thus again allowing full moisture movement. The vertical channels or projections formed in these two cases have the advantage that they check the horizontal flow and accumulation of water across the wall under the influence of the wind.

The finish of various types of sheathing at internal and external angles requires some consideration, both from the point of view of appearance, and of weather resistance. The sheathing can stop at the corner against a vertical member, the joint being midway over a stud to give satisfactory end nailing. It is as well to insert an extra strip of building paper at these points. Fig. 599 gives the finish to an internal angle with shingles. The 44 mm square in the corner is necessary to give the required depth.

Fig. 600 shows an external angle which can be used with any type of timber sheathing. In Fig. 601, the top layers of double shingle walling are brought forward to meet at the angle with butt joints alternating on opposite faces.

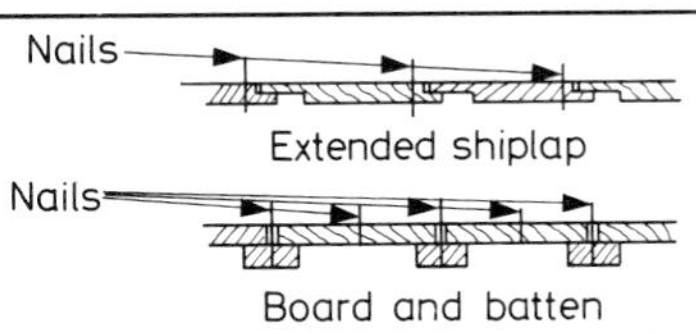

therefore stand clear so that it may dry quickly when the opportunity occurs. This is shown in Fig. 594, which is a section through the lower finishes to horizontal rebated sheathing. This is known as 'novelty siding'.

This section also indicates the precautions needed against damp when only a part of the floor (say in the kitchen) is concrete so that, in order to be at a common level, it must be above the bottom edge of the wall framing. A continuous damp course is then needed which extends from the skirting down to the foundation wall.

Fig. 595 shows a cement rendered or rough-cast finish on expanded metal. The metal should be rust-proof and should be nailed against thin vertical firrings so that the mortar forms a key behind the metal. The rendering should, preferably, extend over the wall so that it can drain as much as possible.

Fig.598. VERTICAL SHEATHING

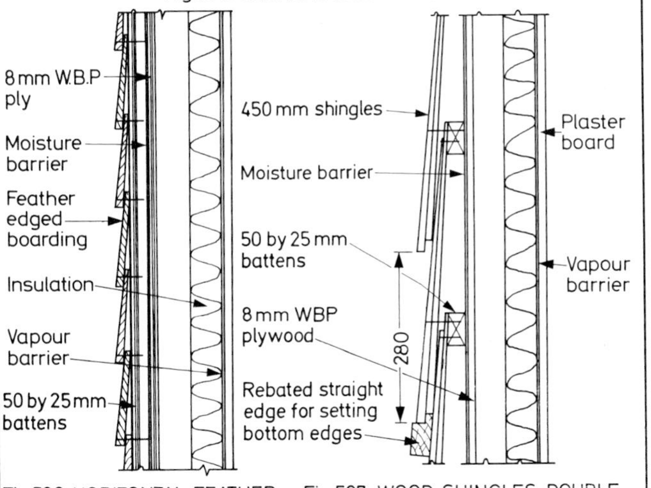

Fig.596. HORIZONTAL FEATHER EDGED BOARDING

Fig.597. WOOD SHINGLES DOUBLE LAYERS - SINGLE LAP

EXTERNAL WALL CLADDING

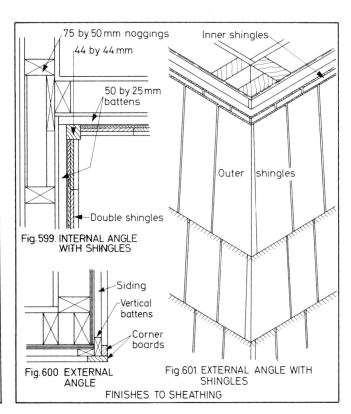

Fig.599. INTERNAL ANGLE WITH SHINGLES

Fig.600 EXTERNAL ANGLE

Fig.601. EXTERNAL ANGLE WITH SHINGLES

FINISHES TO SHEATHING

133

Openings in timber-framed walls

It should be noted that, when an opening is formed in a timber-framed wall, the load carried by the head over is transferred direct, by means of the intermediate studs, to the lintel which should be of a suitable size to take this. The ultimate bending moment and size of lintel required to take it depends upon the size of the opening or the span.

The lintel is commonly formed of two 44 mm thick members, which gives some tolerance to enable both faces of the lintel to be flush with the rest of the framework. Common lintel depths relating to span, accepted as being on the conservative side are:

Lintel depth (mm)	Span (mm)
150	800
200	1200
225	1600
250	2000
300	2400

Double studs are used at the jambs, the inner ones being cut under the lintel to carry its weight direct. Fig. 602 shows a part elevation of a door opening. Note that the outer stud is notched into the sill and sets the width of the door opening.

Partitions may or may not be load-bearing. In the latter case, they only need to have single heads. Where the partition runs parallel to the ceiling or first floor joists at the head, it can be fixed as shown in Fig. 603. Noggings at about 600 mm centres are cut between the joists; they support a longitudinal board which takes the partition head with a nailing either side for the plasterboard ceiling.

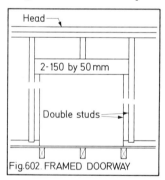

Fig.602. FRAMED DOORWAY

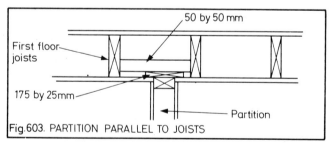

Fig.603. PARTITION PARALLEL TO JOISTS

Most of the services are set within the wall thickness. In Fig. 604, which shows a 75 mm pipe passing through a partition head, the weakening effect on the head of making the necessary large hole is countered by two 75 by 50 mm cleats cut around the pipe and nailed down to the head. Where a greater depth of wall is needed, the pipe may still be concealed if, as in Fig. 605, wider studs are used or the width is made out with firrings.

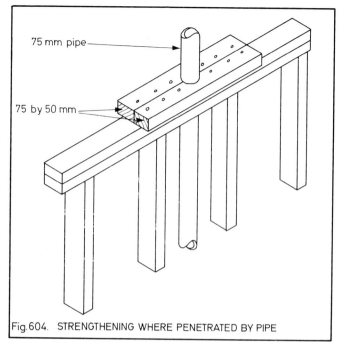

Fig.604. STRENGTHENING WHERE PENETRATED BY PIPE

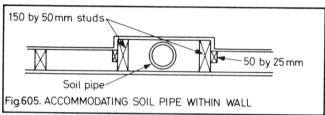

Fig.605. ACCOMMODATING SOIL PIPE WITHIN WALL

As there are no obstructing walls, trimming may be taken completely around chimneys or chimney breasts. 50 mm clearance is provided between the trimmed opening and the chimney. This may be filled with non-combustible materials. When the chimney comes within the length of a timber-frame wall or partition, the brickwork may be extended, on either side if necessary, in a half-brick thickness to give a fixing for the studs at a safe distance from the flue (Fig. 606).

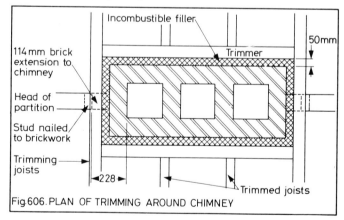

Fig.606. PLAN OF TRIMMING AROUND CHIMNEY

Openings for stairs may be formed in the usual way. The wall string will be carried by the studs of the framing, to give a fixing for the plaster-board or other lining against the string. Battens 75 by 25 mm may be housed flush behind the raking top and bottom edges of the string, as shown in Figs. 607 and 608.

Any weakness due to the cutting away of the studs will be compensated by the additional stiffness provided by the

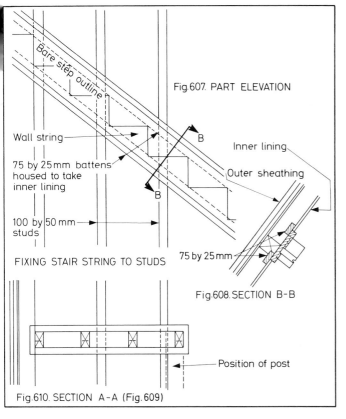

Fig.607. PART ELEVATION

Bare step outline

Wall string

75 by 25mm battens housed to take inner lining

100 by 50mm studs

FIXING STAIR STRING TO STUDS

Inner lining

Outer sheathing

75 by 25mm

Fig.608. SECTION B-B

Position of post

Fig.610. SECTION A-A (Fig.609)

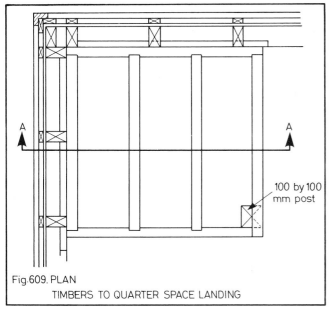

100 by 100 mm post

Fig.609. PLAN
TIMBERS TO QUARTER SPACE LANDING

stairs. Quarter-space landings (Figs. 609 and 610) may be framed up and nailed back to the studs; a board wide enough to provide nailing for the linings both sides of the landing is housed into the studs behind the wall timbers (Fig. 609).

JOINERY FINISHINGS

The joinery forming the finishings of timber-framed buildings may, to a large extent, be of standard design, as fitted into brick or other types of walls, load-bearing or otherwise. However very careful detailing is necessary to ensure that pockets of damp are not allowed to remain permanently within the structure. If dry rot were allowed to take a firm hold, the results could be disastrous.

Close surfaces, which are not hermetically sealed, offer a strong invitation to capillary action. At the same time, they tend to retain the moisture during dry periods, so that regularly recurring downfalls of rain ensure its uninterrupted continuance.

Casement windows

Figs. 611 and 612 are sections through a casement window in the wall to a timber-framed building with 112 mm thick brick veneer facing and a 50 mm cavity. In the horizontal section (Fig. 611), it will be seen that the standard EJMA-type windows have been used; although the width of the jambs has been increased to allow the casement to open under the condition shown. Assuming that dry linings, e.g. 12 mm plaster-board, are used to the insides of the walls, the window recess is most conveniently brought out with timber linings tongued into the jambs.

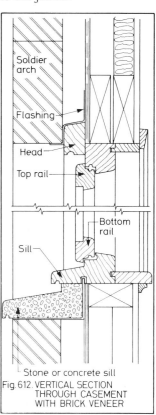

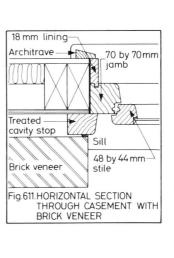

18 mm lining

Architrave

70 by 70mm jamb

Treated cavity stop

Sill

Brick veneer

48 by 44mm stile

Fig.611. HORIZONTAL SECTION THROUGH CASEMENT WITH BRICK VENEER

Soldier arch

Flashing

Head

Top rail

Bottom rail

Sill

Stone or concrete sill

Fig.612. VERTICAL SECTION THROUGH CASEMENT WITH BRICK VENEER

If the cavity was sealed as in normal brick construction, the window would have to be brought forward to cover the tiles, slates or other material used to fill the gap. This is avoided by using a timber cavity-stop finished to agree in appearance, with the general joinery. The stop should be nailed to the plywood linings over the moisture barrier before the brickwork is built. It should be pressure-impregnated with a type of preservative which, when dry, may be painted over. A mastic sealant may be applied in strip form or with a gun to isolate the timber finally from the brickwork.

Fig. 612 is a vertical section through the same window. In this case, the cavity-stop is weathered on the top edge and is covered by metal, or another flashing, taken out under the brick arch. The flashing should be close nailed

135

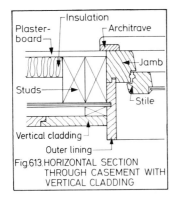

Fig.613. HORIZONTAL SECTION THROUGH CASEMENT WITH VERTICAL CLADDING

a neat all-round boxed appearance, the jamb is tongued into, and finished flush to, the sub-sill (Fig. 614).

Fig. 614 is a vertical section through the window. In this case, water running down the vertical cladding is likely to be plentiful, so it is thrown well clear of the window by a projecting head weather mould with a drip. The moisture barrier should be taken under the rebate in the cladding. The bottom of the cladding, being end grain, will be vulnerable to damp absorption and should be kept above the weathering of the head to allow for rapid drying between storms. The sub-sill is tongued into the frame with a drip which projects over the cladding. It also has a groove to take a narrow strip of sheeting as before which then seals off the main framing.

to the plywood lining and should come under the vapour barrier. To remove the risk of driving rain getting behind the rebate of the top rail, the bottom of the cavity is left square with a convenient drip.

As the sill to the window frame is set well back, an outer or sub-sill of tiles, stone, or pre-cast stone as shown, is advisable to carry the water clear of the wall. The cavity is sealed with metal flashing grooved up into the wood sill against the plywood lining and taken to the outside of the wall. Inner finishings are standard with window boards and bed mould to receive the architraves.

When the outer sheathing is of timber, either vertical or horizontal cladding, or wood shingles, some consideration must be given to the way in which the linings to the opening are taken from the window-frame to mate with the cladding without leaving any weak points in weather resistance. One way of accomplishing this is illustrated in Figs. 613 and 614. In the horizontal section (Fig. 613), outer jamb linings continue out to cover the edges of the vertical cladding; a satisfactory seal is provided by a separate strip of plywood, as on the outer wall sheathing grooved into the lining. In this way, the window may be fixed after the main sheathing has been applied. To give

Sash windows

The use of double hung sliding sash windows, although not as popular as in the past, is still adopted in some cases; particularly in exposed positions where gales could cause damage to hinged open casements.

An example is given in Figs. 615 and 616, where it is assumed that the wall is clad externally with horizontal novelty siding. Assuming normal section sashes, the depth of the boxing to the window must, as a minimum, be the sum of the thicknesses of the sashes, beads and outer lining which is also about the overall depth of the finished wall. As the structural jambs (100 by 50 mm studs) provide a smooth surface, no back linings are necessary, but the 15 mm inner linings must be rebated to come flush with the face of the plaster-board. The outer jamb linings must form a stop to the wall siding; to this end, they are tongued into fairly stout pilasters as shown in Fig. 615. Referring to the vertical section in Fig. 616, the construction of the boxed head follows normal practice; but a head weather mould is used to throw the water clear of the line of jambs and sill.

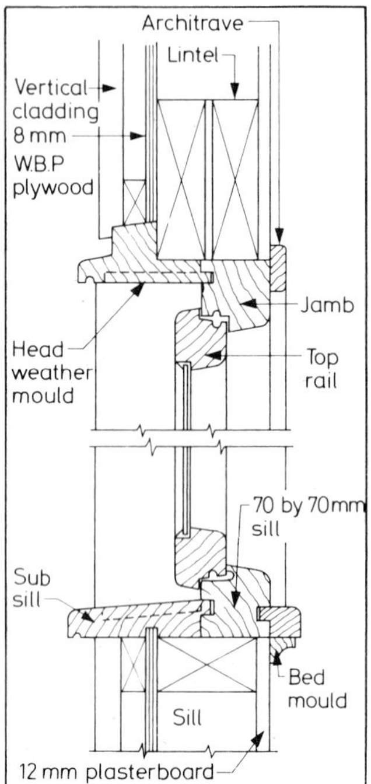

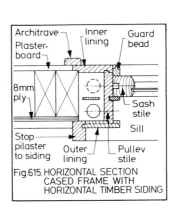

Fig.615. HORIZONTAL SECTION CASED FRAME WITH HORIZONTAL TIMBER SIDING

Fig.614. VERTICAL SECTION THROUGH CASEMENT WITH VERTICAL CLADDING

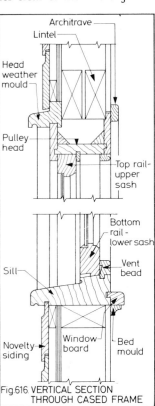

Fig.616 VERTICAL SECTION THROUGH CASED FRAME

The outer lining or casing is housed into the weather-head, which is rebated back to the thickness of the sheeting. The joint should be covered by the moisture barrier. A gap should be left under the bottom edge of the siding which should be bevelled as shown. The boxed frame may be constructed in the normal way, but the sill should be extended to the pilasters. This means that the outer linings should be tongued into the sill with the shoulder on the outside as, also, should be the pilaster.

External doors

External doors have additional problems with the threshold, as well as with the jambs and head. Figs. 617 and 618 are sectional details through a doorway in a timber-framed building with horizontal siding. It is shown with a small weather-head, or pediment, to give some protection to the person opening the door. In the vertical section, it is seen that the weather-head is of boxed construction with a large cornice mould to the fascia. The ends are returned to an outline following the section.

The weathered top is covered with metal flashing, which should be taken up under the timber siding and moisture barrier and carried down to form a drip. The frieze board under the soffit should be thick enough, so that it can be framed up flush with a stout pilaster (shown in horizontal section in Fig. 618) which will form stops to the horizontal timber sidings.

It is important that the bottom of the door, together with the threshold, should be so designed as to eliminate the possibility of rain being driven in over the floor. This is achieved in the example given by a weather-mould, with a suitable drip tongued into the bottom rail of the door and extending over a sunken weathered edge to the threshold. The threshold should be of hardwood, say oak

or teak. A brass strip could, with advantage, be recessed into the threshold at the exposed angle, to take the wear. The weather-mould should be of sufficient width to allow the water to drip clear of the edge of the sinking.

The projection of the threshold has to be considered in relation to the main walls of the building. Water must not be allowed to reach the plate carrying the floor joist. To this end, the sill weather-board is continued along under the door. The riser to the threshold step is tongued into the threshold and cut down on to it, as shown in Figs. 617 and 618.

A cleat nailed inside the heading joist, opposite the doorway, will take the joint between the threshold and the floorboards. Internal doors do not offer much of a problem as the vertical studding at the jambs will give a more accurate fixing than is usually available in, say, brick partitions. Fig. 619 shows the usual cheap linings, with stops planted after the door has been hung, to accommodate any slight irregularities in the door. Ample nailing is available for linings, architrave or band mould, etc. No rough grounds are needed, unless the studding is lathed and plastered (which is unlikely).

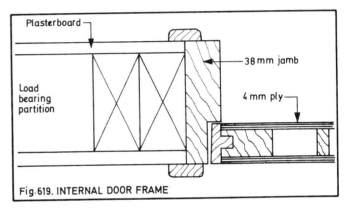

Fig.619. INTERNAL DOOR FRAME

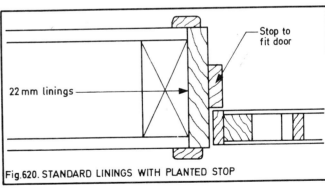

Fig.620. STANDARD LININGS WITH PLANTED STOP

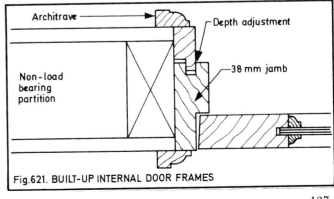

Fig.621. BUILT-UP INTERNAL DOOR FRAMES

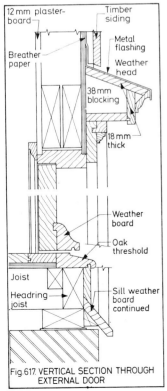

Fig.617. VERTICAL SECTION THROUGH EXTERNAL DOOR

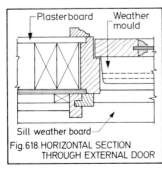

Fig.618. HORIZONTAL SECTION THROUGH EXTERNAL DOOR

Fig. 620 shows a solid door-frame rebated one edge to take the door in a load-bearing partition, where double studding is advisable to carry the load imposed on the lintel over. In Fig. 621, the jamb is built up in two parts with a dummy rebate at the back edge. The advantage of this is that the width of the jamb can be adjusted to suit any slight variation in the overall depth of the partition framing. It is suitable for frames with pre-hung doors but the door opening should be a little wider so that, when the jamb to the hanging stile is fixed, the other jamb can be adjusted to fit the closing stile.

Assembly order of timber-framed buildings

As stated previously, the preliminary assembly routine is to construct the foundations and brickwork to fireplaces, etc., up to first-floor, and to lay floor-joists with sub-flooring of plywood or diagonal boarding. The construction then continues to assemble and to erect walls and partitions, applying external ply sheathing only with the protective moisture barrier. This is followed by the first-floor, first-floor walls and partitions, and then the roof and external doors and windows.

The building is now weather protected and work on the interior can proceed as convenient while the exterior can be continued as weather permits. This will include exterior cladding, erection of half-brick walls (when the building is brick-veneered), applying other exterior finishes, painting, etc.

Inside the building, pipes and cables to the plumbing, electric and other services can be installed while the framework is still open and before the vapour barriers and internal linings are applied. Finally the internal finishings are installed, after the linings; which include finished floors, joinery and fitments, skirting, architraves, etc.

When the external finish is of brickwork, there will not be more than the normal amount of timber showing and needing protection. Where the external cladding is of timber, large areas would need to be treated.

Where the cladding is Western Red Cedar, this is naturally decay resistant, but it will weather to a silver grey colour which might be considered drab by some standards, although it can be relieved by the paintwork. The application of a protective clear varnish has not, in the past, proved to be successful. The exposed surfaces of the bare cedar will become dirty, at which stage they can be cleaned by the application of a sodium hypochlorite solution (bleach) or oxalic acid. This must be washed off afterwards with clean water. The surface may then be allowed to weather again, or be treated with a colour preservative.

APPENDIX A

The symbols used in structural calculations over many years and including those in the new superceded CP 112 1971 have been replaced in part in BS 5268 by others in general accordance with ISO 3898 published by The International Organization for Standardization. To avoid confusion the new terms and old together with their oral description (in the case of Greek letters).

New symbol	Name	Previous symbol	Representing
a			distance
A			area
b		b	least dimension in beam section, or post section, or tie section
d			diameter
E		E	modulus of elasticity
F		W	force or load
f			unit load
h		d	depth of beam or greatest dimension of a cross section member
i		r	radius of gyration
K		K	modification factor (generally to stress)
L		L	length or span
M		BM	bending moment
n		n	number
r			radius of curvature
t			thickness of lamination
α	alpha		angle between direction of load and direction of grain
η	eta		eccentricity factor
θ	theta		angle between longitudinal axis of a member and a connector axis
λ	lambda		slenderness ratio
σ	sigma		stress
τ	tau		shear stress
Z			section modulus
ω	omega		moisture content

The subscripts used are

(a) *Type of force stress etc.*

c		c	compression
m		f	bending
t		t	tension

(b) *Significance*

a	applied
adm	permissible
e	effective
mean	arithmetic mean

(c) *Geometry*

apex		apex
r		radial
tang		tangential
$\parallel$		parallel to the grain
$\perp$		perpendicular to the grain
α	alpha	angle

Subscripts may be omitted when the context within which the symbols are used are unambiguous with the exception of the modification factor **K**. e.g. if a joist supporting decking to a concrete slab has to be designed for safe span when the timber is of known section and of a known strength class then the basic stress value taken from the tables will represent the safe fibre stress in bending when used in a dry condition (MC 18%) and under long term loading, whereas the timber used for formwork is likely to have a high moisture content and subject to a short term loading. The stress value written in full therefore will be $\sigma_{m,adm}$,K_2,K_3, where K_2 = modification factor for wet timber and K_3 is the modification factor for short term loading.

Assuming now that the basic bending stress = 6.2 N/mm^2 and that $K_2 = 0.8$ and $K_3 = 1.5$ then $\sigma_{m,adm\parallel}K_2K_3 = 6.2 \times 0.8 \times 1.5 = 7.4$, then within the context of the requirements stated above σ may be taken to represent 7.4 N/mm^2.

Graduation of timber species into strength classes

The stress grading of timber into GS, SS, M 50 and M 75 as carried out separately in the different species according to BS 4978, 1973 gave unit stress which varied according to the species of timber so that the SS value in say Western Red Cedar was less than GS grading in Douglas Fir. In the new BS 5268 part 2 this anomoly has been or can be overcome by placing the various species in different "STRENGTH CLASSES" for the various stresses and E values. For instance Table no 8 in the standard gives 5.3 N/mm^2 for bending in strength class SC3. Table no 3 gives GS grade for parana pine, pitch pine and douglas fir and GS/M50 for redwood. According to Table no 9 these have strength in bending respectively of 6.4, 7.4, 6.6, and 6.4 N/mm^2 so the values in the strength class table are rounded down to give a small safety factor.

Table no 8 in the standard gives the permissable stresses, E values and approximate densities of nine specified strength classes SC6 to SC9 are only likely to be met in a dense hardwood.

INDEX